DOGS THAT EAT GODS

The God Hallucination,
by **J D Fleming**

How the human brain created myth and religion, taking credit from the powers of the Universe.

A book about the Universe, Evolution, the Human Brain, Religion and the Future. A non-biblical account of how we got here, and where we are headed.

Copyright © JD Fleming 2025

The moral right of this author has been asserted.

All rights reserved.

No part of this publication may be reproduced, stored in a retrieval system, or transmitted, in any form or by any means, without the prior permission in writing of the publisher, nor be otherwise circulated in any form of binding or cover other than that in which it is published and without a similar condition including this condition being imposed on the subsequent purchaser.

Cover design by Brian Grogan

Editing, typesetting and publishing by UK Book Publishing

www.ukbookpublishing.com

ISBN: 978-1-917329-64-4

I dedicate this book to my wife *Lesley* who has been so supportive during its writing and the other 40 years, my mother and father for giving me choice and freedom of thought, and my daughter Bex for demonstrating compassion and intelligence through her ideals.

Special mention to *Rachel Z* and *Brian G* who helped me along the journey.

* The book title 'Dogs that Eat Gods', is a reference to how humans have changed Gods throughout history, often as an outcome of wars … the dogs of war. The dogs that eat Gods.

Warning
- To read this book you need to be:

1. Of above average intelligence, an IQ of 110 or above, with a supporting level of emotional intelligence is recommended, or you may be wasting time and money. The primary aim of the book is to be informative and challenging; it is not primarily written for economic and commercial gain. A recommended IQ of 110 effectively disqualifies 66% of the population of developed economies, a much higher proportion of those living in underdeveloped economies.

2. You need to be open minded with an ability to, at least, consider new points of view and factual findings that may challenge your current view of the Universe and planet you inhabit.

3. If steeped in dogma, especially religious dogma, this book may make you angry and upset. Do not read on, it is not the intention of this book. If not prepared to at least consider new or different perspectives, stop reading now.

Note: It contains information which science is quickly turning into knowledge. I am envious that in around 400 pages, you will gain a level of knowledge it has taken me years to establish. However, that is what non-fiction writing aims to achieve.

Objectives for this book.

In my experience, books on specific subjects, written by experts, are long and detailed, deterring all but highly interested parties, and those that can comprehend great detail or high complexity. This book instead aims to provide a non-detailed account, with minimal complexity, of the areas that we should understand, at least in part, to gain perspective on our world. I have endeavoured to link these into an evidenced story, with knowledge in parts, theorised in others using available information and balance of probability.

It further aims to:

1. Challenge fundamental and possibly outdated human beliefs, with emerging knowledge.

2. **Take the reader on a journey through history and discovery of our species, the huge forces that conspired and unknowingly cooperated to produce us, against huge odds. To challenge readers' minds.**

3. Give an insight into what humans have achieved with their exceptional brain power, yet also highlight how easily influenced and naïve the

human brain can be, and conclude with how it may all end for humankind.

4. Convey a personal record of a six year-old-boy who learned and changed slowly but surely, he eventually changed his perspective and views in light of emerging information, born of research and modern learning, thanks to science. Many individuals struggle to reconsider or refresh their viewpoints in light of overwhelming evidence; I want people to acknowledge, 'this author did not fall into that category'.

5. Cover the areas of emerging modern learning and information that should lead us all to at least a small change in perspective.

6. **Take the reader on a journey, every human has the right to experience or have explained.** To give an accessible, non-detailed reference, in a book which may have been written, at least in part, centuries ago, had the information been available and death threats from the ignorant, left aside.

7. Demonstrate that humans aren't generally prepared to accept new knowledge and will seek to influence future generations, with guesses disguised as truth, by reinforcement and exclusion of alternative views. My mother and father, grandparents and forebears had an excuse

– information was not available, compulsory indoctrination was common. My son, daughter, and future grandchildren must recognise that they have no excuse; perspectives need to evolve with new information and knowledge. We should remain open to re-evaluating our beliefs and what we consider givens, even if we decide to stick with our original views; otherwise, our opinions risk becoming inflexible and irrelevant.

CONTENTS
—

Foreword 1
About a Boy who changed, the effect of knowledge on perspective, the wonders of learning from new discoveries. Your story, the story of every human

Chapter 1 - Evolution 28
A brief run through of the formation of Earth and early life, how it persisted in a hostile environment against all odds

- **Formation of Earth**   29
- **Single Planet, Single Cell**   36

Chapter 2 - The Ages of Earth 39
The rise of multicellular life, its battle to survive in spite of five mass extinctions, decimating the majority of life on Earth.

- **Multicellular Earth - The Next Major Transformation**   43
- **Oxygen: Rapid Progress and Disaster**   48

Chapter 3 – What is Life 63
A simplified view of the chemistry of life, metabolism, respiration, DNA and replication. Natural selection, adaption and survival by changing DNA. Volume of survival, a 3.8 billion years' unbroken chain.

- **Instincts and Natural Selection**   72
- **The Volume of Survival**   77
- **The Beat of Life Goes On**   83

Chapter 4 – Hominins 89
The rise of humankind and its amazing brain. The genius of the human brain in discovery and invention. The vulnerability of the human brain expressed in superstition, insecurity, need to belong for comfort and through compliance. The conceiving and spread of religion. The naive influenced human brain and its astonishing capability.

- **Humans and Bacteria**   91
- **Humans and their Brains**   95
- **Nutrition Driven Faith**   110
- **The Influenced and Manipulated Human Brain**   117
- **Psychological Change**   123
- **The human brain, a thing of wonder, a thing of naivety**   124
- **The Human Brain and God(s)**   134
- **Battles for the Human Mind**   140

Chapter 5 - Realisation and Consciousness 163
The human brain experiencing consciousness, the need to belong, exploitation by humankind and the effect of the information age. The conditioning and manipulation of the human brain.

- Human Brain 'Superiority' — 190
- Conceptualising God — 192
- Before the Beginning — 198

Chapter 6 - Universe — 207
A brief description of the Universe and its wonders. Hubble and James Webb, a window to spacetime. Life beyond our solar system. Humans a 'tiny widow in time'.

Chapter 7 - Incredible Scale — 216
A summary of the scale of power of the Universe, how many stars, life and death of stars. How long stars live and die, and their legacy. The creation of elements, essential for life, within stars. We are stardust.

- The Universe (continued) — 219
- How Many Stars? — 222
- Stardust not Golden until Death — 223
- Galaxy Super Power — 227
- Routine Death of Stars — 230
- Are we alone? — 240
- Narrowing the Search — 243

Chapter 8 – The Universe's Candidates for known Gods 257
A brief summary of the superpowers all around us and their justification for being considered Gods

- Candidates for Gods of the Universe, black holes, quasars, electromagnetism, strong nuclear force, time, gravity — 262
- The Lesser Powered Known Gods — 283
- Candidates for Gods of our galaxy – the Milky Way — 285
- Candidates for Gods of our Solar system, the Sun, Jupiter — 286
- Candidates for Gods of Earth, the moon, oceans and seas, natural selection and evolution, humankind, bacteria and microbes — 298
- Summary — 298

Chapter 9 – What Lies Beyond 301
What are the dangers lurking in the Universe to our planet and our species.

- Destiny, Certain and Irreversible — 304
- An Asteroid or Meteor — 306
- The Moon Moving away from Earth — 308
- Loss of Earth's Magnetic Field — 308
- Solar Blasts — 309
- Gravitational Waves — 310
- Quasars — 311
- Black Holes — 312

- Supernova 313
- Self-inflicted Extinctions – Earth 316
- Long Term – Can Hmans find a new home, before Disaster Strikes? 318

Chapter 10 – Humankind 321
Still evolving and destined for something better. Human co-existence and the role of natural selection. How humankind will diversify and progress, before we come to an end.

- Forcing the Pace 343
- Religion and Early use of Human Advantage over Humans 349
- Dominant yet temporary 353

Acknowledgements and Inspirations 379

DOGS THAT EAT GODS

The God Hallucination,
by **J D Fleming**

FOREWORD

—

'Isn't it enough to see that a garden is beautiful without having to believe that there are fairies at the bottom of it?'

DOUGLAS ADAMS,
Hitchhiker's Guide to the Galaxy'.

*

I developed a trusting nature from an early age, similar to many others. I was born in the late 1950s with memories from childhood, mainly, almost exclusively, in the sixties. A different world from today. I believe that those 60-plus years have seen more change to humanity than any 60 years period since Homo sapiens have existed. I anticipate the next 60 years will bring about an even more rapid pace of development and human accomplishment.

Over its 4.6 billion-years existence, the Earth has undergone significant changes, so in the grand scheme of time 60 years feels like a moment; nonetheless, that brief period has witnessed remarkable transformations.

DOGS THAT EAT GODS

Life was so much more straightforward in centuries gone past: a parent, teacher or elder told a young person something, it was believed. It was fact that needed no confirmation, and usually taken to the grave. There was no reason not to believe, or to be more accurate, there was no accessible alternative or conflicting information. It served as a definitive doctrine. In my childhood, I believed the Earth was only 6,600 years old. James Ussher, the 17th century Archbishop of Armagh, in Ireland, calculated this. How it has aged in my lifetime.

When I was a young boy, horse and carts delivered our coal which was burned without guilt as we watched thick smoke rise as the coal had been rain dampened in transit. Lifted on the backs of men, the coal was delivered to our bunker and burned over the next week or two.

I remember lying by that fire the day the President of the United States, John Fitzgerald Kennedy, was assassinated, my concern running to whether there was enough coal in the bucket aside the fire or whether I, or my sister, would have to go out to the bunker on a cold night to get more. That experience was my earliest memory of perspective.

When I realised how shocked and upset my mother was at the news about the President, in that moment it suddenly became more important than the fire. When TV was shut down for the night as a mark of respect, even at six years old, I got a clear message: something really important had taken place.

FOREWORD

Perspective dominates our lives as humans, wars, religions, politics and much more, born out of experience and hopefully learning. Most of us are familiar with the term 'one-man's terrorist is another man's freedom fighter', perspective formed through exposure to life from different experiences. The circumstances of our birth, our inherent qualities, and our early life experiences significantly influence who we grow to be, allegiance to a family, country or religion for example.

For perspective to be accurate, it must be linked to knowledge and truth, not purely driven by information, emotions of hatred, love or biased, as so many human perspectives are in our contemporary world; we cannot truly achieve accurate perspective without truth. I have discovered that governments, churches, corporations, and even close relationships often conceal the truth or tell lies. Some of these can have good intentions and may be harmless, but they can also cause damage and often lack reflection regarding consequences. I have learned that truth can only be established through knowledge, not just information. The greatest lesson I've learned in my life.

The greatest lie of all, if it is not true, may be the existence of God, as conceptualised by humans. This question has long challenged humanity, affecting their thoughts, emotions, loyalties, and lives. Various interpretations of God exist, have existed or will exist, including those that have faded with time and those yet to emerge.

Once the human brain is 'signed up' to a concept of a God as described in its childhood, it has, in many cases, gone through an irreversible process. The degree of which appears to be commensurate with the level of exposure to information and the consequences of not following religious doctrine, within different cultures.[1]

Humans have a tendency to side with the familiar rather than stretch our minds and think what we know could be wrong, or at least lack perspective. How does new information change perspective? From my experience, adult humans often resist accepting new information as knowledge when it challenges their existing beliefs. Few humans want to admit they have been or may have been wrong, over considerable past time. I plan to delve deeper into this area when I tackle topics related to the human brain, which others may consider to be controversial or even forbidden.

Change of mind, beliefs, views happened slowly, over time, for me. Over the past 60 years, humans have made numerous discoveries and inventions, significantly accelerating during that period. The years between 1960 and 1990 offered an amazing experience for lucky individuals who lived during that era.

Television sets, washing machines, cars, home ownership, vacations abroad, landline telephones in most residences, heart transplant procedures, computers in homes, widespread mobile phone access, lunar expeditions, the successful eradication of smallpox, and the development of cures for various other health issues.

As we observe deeper into space, we also look further back in time, revealing an older narrative. Today, the images departing from Earth will reach that destination in 2.5 billion years, depicting a lush planet bustling with life. By the time they reach their destination, scientists predict planet Earth will lack any signs of life, all multicellular organisms are expected to have become extinct due to the Sun's progression towards becoming a red giant, ultimately leading to its demise, and that of our solar system, in about 5 billion years.

A dying solar system, a lifeless Earth, possibly with unseen single cell creatures, and an ever growing, deep orange Sun expanding as our solar system grows old. Any life remaining on Earth will likely be extinct, or underground away from the scorching heat and will be heading for extinction.

The same is true for humans looking for life in other galaxies. Andromeda, the closest large galaxy to us, lies roughly 2.5 million light years from Earth, while the most distant galaxies are located tens of billions of light years away.

In 2.5 billion years our Sun may still have more than 20 percent of its life to go, but will be doomed to implode when gravity becomes a stronger force than the forces of our star, as its fuel supply exhausts.

The future of the Earth, and its past, is in the stars, not in a horoscopes sense but in a very real sense. Scientists

FOREWORD

But when compared to the last 30 years, progress in that 30-year window (1960-1990) now appears slow. During the last 30 years, scientists decoded the human genome and introduced artificial intelligence. We have created compact phones that can capture images, record audio messages, and offer nearly limitless information. But top of my list is the mechanisms that help us look across the Universe and see our past and our future.

Hubble, James Webb and other high-resolution, multi-dimensional telescopes, now enable us to look across space and back in time. They serve as 'time machines' and are also very costly pieces of equipment.

Observing a light turned on across a room requires just an instant for it to arrive at our eyes, travelling at approximately 186,000 miles per second, about 5.88 trillion miles per year. By the time we register sunlight, more than eight minutes will have passed from light leaving the Sun, illustrating a moment of looking backwards in time. When we gaze towards the centre of our galaxy, we witness an event from 26,000 years ago because that's how long it takes for light to reach us from that distance.

A civilization located 2.5 billion light years away from Earth, in a different galaxy, would currently see images of our planet as it appeared 2.5 billion years ago, when Earth hosted only unicellular life forms and appeared to be active environmentally, yet lifeless. Single cell creatures, the only inhabitants with no sign of the multicellular, complex life yet to come.

FOREWORD

using data from Hubble and James Webb telescopes have already observed the birth of solar systems and the deaths of others, much like our own. Medium-sized stars some with expected lifecycles of around ten billion years.

We may never know whether life exists or not, but we can be certain that the cycles of life and death among solar systems have occurred trillions of times and will continue this process for tens of billions of years, or longer. We can observe how our Sun was born, how it gave existence to planets, life, and how it will, one day, destroy us all, unless we have the science that enables humans to escape our solar system.

I didn't have this information when I was six or even at 40 years of age, but I feel very fortunate to know it now. In short, my perspective has been changed and the Universe somewhere along the way aged very quickly, in my mind and in reality. In less than 60 years, I have moved from observations of the Universe being 6,600 years old to understanding that the Universe is a little under 13.8 billion years old. Transitioning from faith in religion, to a belief in science, has profoundly changed my understanding of the world, surpassing my childhood perceptions and expectations.

This book aims to provide people with a new perspective on the incredible journey that the Universe, our Galaxy, Solar System, Earth, and its species have experienced. The narrative of our solar system and planet unfolds over a long timeline, brimming with both marvel and sorrow.

I will also briefly discuss, the remarkable story of evolution and how it led to the development of the human brain over 3.8 billion years. Starting with possibly the least interesting two chapters of a book may not be a good plan to encourage readers to finish this book; it is, however, crucial to set an accurate context. Knowing how we got here is important context we should all be aware of.

I find it astonishing that so many people neglect the opportunity to explore our Universe, now that we have the means to do so. For millennia, humans have pondered what lies beyond the farthest stars visible to the naked eye, or beyond the capability of Earth-based telescopes grappling with the distortions of Earth's atmosphere.

Religion offered theories in the form of opinions, about what exists beyond our planet, but the Hubble and James Webb telescopes revealed a nearly infinite number of galaxies and stars in the Universe. A massive estimate could be the closest we may ever get to knowing its full and wonderous extent.

The powerful influences of the Universe remain overlooked by many human minds. Most people remain unaware of the forces that drive the Universe. The promotion of God(s) in human thought has stripped away mystery, providing an unfounded explanation that leaves no room for alternative narratives. Inseparable ideas have taken root in human minds over thousands of years, becoming widely accepted as truths. Powers that are unlikely to exist outside the human brain, viewed as highly significant; actual proven powers of the Universe, ignored.

The contributors of the Christian bible presented the story of creation, similar to other religious texts. This usually contains information about an invisible, omnipotent God. What would the story of creation look like today, using science, known facts, theoretical or simulated evidence using available data, or where we don't have sufficient information, simulate or develop a feasible theory, using balance of probability? Many scientific theories eventually achieve confirmation as facts, such as the existence of gravitational waves that traverse spacetime, a notion proposed by Einstein. Gravitational waves are ripples in spacetime; the more easily detectable are caused by cataclysmic events such as colliding black holes, colliding neutron stars or massive stars going supernovae.

When translating a religious explanation into a scientific one, creation can represent the Universe's story, encompassing the 'Big Bang' and the evolution of stars and galaxies. Genesis of humankind would be the story of evolution from single cell to the ascent of modern humans.

The story of human morality, good and evil would be the story of the human brain and how it is set apart from animals, how it has created morality, and with it, different standards for itself. Armageddon and the 'last day' would be the story of the end of humanity, followed by the end of Earth and our Solar System, then the end of the Universe itself. We can only make estimates based on the information we possess, but this is probably more accurate than the four horsemen of the apocalypse or other ancient religious predictions, or metaphors, of the far future.

Earlier, I outlined the objectives of this book. This reflects my understanding of past events and future possibilities, influenced by insights from highly intelligent individuals. The context of the superpowers in the Universe highlights their crucial involvement in the creation of matter, stars, planets, and consequently life. Modern humans with an interest, deserve to know so they can arrive at their own conclusions, having first absorbed available information and a range of interpretations.

When we set aside religious texts, which were written by humans such a long time ago that they lacked the information we have today, we can gain a clearer perspective on the origins of the Universe and evolution. A narrative that captures the amazing reality of Earth and humanity's fortune, reflecting on the odds we overcame to exist in present day. Life and evolution have travelled a journey spanning 3.8 billion years, and it continues to unfold.

For the most part, Jupiter serves as the 'big brother' by protecting us from powerful forces and large objects colliding in space; recent science has shown that it can

FOREWORD

also sling shot smaller objects towards Earth, but it is on balance a positive influence.

The moon plays a vital role, as a small but important sibling. How life on Earth is thought to have started. Over the course of 300 million years, life sought to gain a foothold and develop reproductive capabilities. Single-celled organisms endured a challenging battle that lasted over two billion years before multicellular life was able to persist. The combination of nuclei plus energy producers, photosynthesis and mitochondria, created a favourable environment for development of multicellular organisms.

Over the past five hundred million years, five major extinction events occurred, each erasing at least 75% of Earth's life forms. The Earth shows remarkable durability and powers of recovery.

If humanity were to make itself extinct and cause great destruction to Earth's life, it would merely be a temporary setback. This event would be unlikely to match the mass extinction that occurred more than 250 million years ago, during which 96% of life on Earth was made extinct. The Earth recovered, but we will inevitably face extinction-level disasters in the future; the only unknowns are their timing and magnitude. The Earth and nature don't grieve for a lost species; they simply begin anew.

As you read my book, remember that the story is yours just as much as it is mine and that of our world. This story encompasses every living being, including all that have existed and all that will come to exist on our thriving then dying planet and beyond.

I have attempted to eliminate human emotions from our story while explaining the influence of those emotions on its development. In my view religious practice engages the human emotional brain to a far greater extent than the logical, practical brain; this possibly exploits the fragilities of the human brain. An understanding of our ancestors, prior to religion, helps us grasp who we are, our feelings, and our evolutionary role, our place in the Universe.

Our existence in the present moment prompts us to believe that we play a greater role in the history of our planet than we truly do. A million years ago, Homo sapiens didn't exist. In a further million years humanity, as we know it, is unlikely to exist, yet our planet will continue its narrative of evolution, giving rise to new species of plants, animals, and fungi, continuing the processes of natural selection.

Tectonic plates will still be active, asteroid impacts and extinctions will still be unpredictable in timing, yet certain in delivery. Our Universe will unfold countless trillions of chapters, long after humanity and our solar system fade into a 'once upon a time'.

Long after our solar system and planet have met their demise, a new story will unfold somewhere in our galaxy and perhaps even beyond. The remaining debris of our solar system will migrate to new systems and be reused as our solar system reused materials from long dead stars. Our bodies consist of chemistry that originated in long-dead stars, and one day, it may reunite with new stars or drift in eternity.

Humans have played a minor role in the overall story of our planet, but this role holds significant importance to us. Our lives have a hierarchy of importance where we consider ourselves, family, and friends as fundamental; others may choose to rank God or their country first. This will typically taper to humanity, animals, environment and so on.

The people and activities we regularly engage with usually take precedence, yet we're also interested in broader issues that impact our lives and our planet. From the Universe's perspective, influences appear quite different, humans playing little part, and this will become apparent as we delve into the topic.

You will receive a brief introduction to the main forces that shape our Universe, which include electromagnetism, gravity, nuclear fusion, spacetime, black holes with immense mass, and stars alongside their planets and moons. Science has demonstrated the existence of dark energy and dark matter through the laws of physics, even though we do not fully understand their nature or mechanisms. For readers looking to delve deeper, this book acts as an appetiser, while a vast array of information stands ready for exploration. Make it your knowledge.

If you haven't already been made aware, you will discover immense powers which exist in our Universe. Powers that can release more energy in a single second than our Sun will release in its 10 billion years lifespan. Objects invisible to the human eye capable of making time almost stand still, others which can destroy anything in their path for thousands of light years of distance including stars and planets.

On Earth, numerous species have thrived for billions of years, far surpassing the existence of humans and likely to be continuing their presence long after the last human has perished.

During my research for this book, I embarked on a journey of discovery that I now want to share. While I don't expect you to accept every word as truth, I urge you to remain open-minded and look for additional information from different sources to deepen your understanding, even if that understanding contradicts mine. Difference and challenge are healthy regardless of what religion may have decreed.

I have attempted to cross-reference and link the findings of the writers and scientists I reference, I have developed opinions and conclusions by synthesising their extensive knowledge and expertise. I have become a very different me from that of my childhood of information propaganda.

FOREWORD

As I have grown older, I have accumulated knowledge that shapes my opinions; until someone presents me with 'definitive' information that transforms my perspective, I will continue to maintain my latest perspective. I embrace the possibility of changing my views, something I believe everyone should be receptive to, even if it hurts to place previous 'knowledge' in the past.

Humans possess the cognitive ability to comprehend the information presented above and throughout this book. So far, to the level of our knowledge, no other species on Earth has displayed this degree of understanding; indeed, humans only developed such comprehension relatively recently in our history.

Despite our size compared to the enormous expanse of the Universe, humans uniquely possess the ability to truly admire its magnificence of the Universe and planet Earth. In spite of this, I predict many humans would fail to grasp the issues considered as well as the findings of this book. A situation that will gradually change over time, as humankind develops.

We exist in a time of learning in which the internet, libraries, television, and other media offer an abundance of information. I want to take you on a journey that presents a different viewpoint on religious scripture and dogma, examining the path that led us to today, and its implications. This encompasses four crucial areas that everyone has the right to assess, regardless of their agreement or beliefs, during their brief time in life:

The Universe, the evolution of life on Earth, the emergence of Homo sapiens and their cognitive abilities, as well as potential – I would say certain – future developments. On our journey, we will likely encounter contradictions between our narrative and several religious writings. Your mind belongs to you, not to me, the church, or the government; you have the power to determine your beliefs.

Your brain holds the jewel of evolution; nothing on Earth compares to its extraordinary abilities. Encourage a focus on knowledge, rather than just surface-level information, wherever achievable. Knowledge refers to information that has been affirmed as correct or reasonably accurate by reliable evidence.

In this book, I might revisit certain topics from various perspectives. The discoveries made by humans interrelate closely with the elements of the Universe, intricately bound within the concepts of space, time, and evolution.

Various forces have contributed to the growth of the human brain and consciousness, many of which we have yet to comprehend fully. Reinforcing significant elements of the text, and even repeating them, will enhance learning and memory recall. I could have made this book slightly shorter by not repeating knowledge, but repetition increases learning, so some key points are re-stated at appropriate stages within this book.

Science, like evolution, is a journey, not a destination. The essence of our being is rooted in ancient powers that existed in the Universe prior to our solar system existing. We took the story up at a later date, when we were born with a realisation which fed off knowledge, as we grew.

Our identity today stems from a combination of our DNA, instinct ingrained in us before birth, and the learning and knowledge we acquire after entering the world. Before our realisation of existence, time had no meaning for us prior to birth, we simply did not exist. We have no memory of it.

We have all heard stories recounted by people believing they have lived a previous life. Yet they would have been a different being, with different views, parents, environments; in short, it couldn't have been them. Could it have been consciousness or realisation passed through DNA causing a chemical image in their brain? We will explore the incredible unconscious intelligence of DNA, nature's most incredible currency.

This book does not follow a chronological order; it begins 4.6 billion years ago but later shifts back to almost 13.8 billion years ago. If life on Earth had not evolved, we would not have attained the present point in our development. The ability of the human brain to explore the Universe. The story of how we arrived here, as accurately as science can get it, we all deserve to hear it.

Choosing to reject science in favour of alternative narratives about humanity's development demonstrates an ignorance rooted in the refusal to acknowledge new discoveries and insights. It's inexcusable not to allow our children the opportunity to develop their own perspectives in life, as a potential alternative to ancient dogma.

Many continue to teach children about Noah's ark and the garden of Eden with removal of ribs and evil serpents, as fact; this, 140 years after the death of Charles Darwin and his amazing work on evolution and natural selection.

Yet some modern historians attempt to airbrush elements of events such as Nazism and Slavery, lessons our children must learn from. We are all aware of the heinous crime of slavery, in transporting people from Africa to the USA, which was a shameful period of our history, yet how many know about North African slavers taking white Europeans captive as slaves. Both very wrong, volume the only difference. It simply isn't discussed because it does not fit the chosen narrative.

When we reach a phenomenon which we can't fully explain, it is simply not good enough to claim an invisible being was responsible and then convince children of this narrative. Many wondrous discoveries have been made rather than accepting a bland 'fits all' explanation. Many great scientists have researched new fields, which were believed to contradict religious teaching, and have come out alive on the other side of the inherent danger they placed themselves in, and been proven correct. Others were not so lucky.

Through evolution and natural selection, humans emerged with a distinct brain structure, particularly emphasising the extra dimension provided by the neo-cortex, frontal lobe. It is difficult to overstate the effect this few pounds of flesh, has made to life on our planet. Composed of atoms and molecules arranged in specific patterns, which ultimately form cells. I plan to examine that topic more thoroughly further into the book.

Our Universe emerged in an instant, then developed over billions of years, following the Big Bang – not a six-day blitz with a day off. Whilst scientists believe that everything began nearly 13.8 billion years ago, for me, it all started shortly after April 24th, 1990, when the Hubble space telescope was launched. I worked back in time from that point. That is, a few years after 1990 when NASA managed to realign faulty mirrors that were making all images blurred. I recommend watching the 50-minute video entitled 'The Invisible Universe Revealed' by NOVA, which you can find online. Fascinating.

A whole new world became apparent, a level of vastness that our ancestors could never have imagined. Following Hubble's launch, NASA examined a section of space no thicker than a drinking straw, uncovering 10,000 points of light. Every point of light revealed itself to be a Galaxy, and each Galaxy encompassed an average of hundreds of billions of stars.

Scientists determined that when applied to the entire night sky, the estimate suggests there are at least 2×10^{22} stars, which outnumbers the grains of sand on all of Earth's beaches. More recently, NASA has estimated that number could be as high as a septillion stars, 1×10^{24} in our universe, 50 times more than the NOVA estimate.

The Universe existed for more than nine billion years before Earth formed. Prior to the Hubble space telescope, we lacked accurate knowledge regarding the Universe's size and age. The minimum estimated size of the Universe is difficult to comprehend. The realisation and then exploration of which, was made possible by evolution natural selection, its latest 'jewel in its crown' of evolution, the human brain. For 3.8 billion years, life on Earth has remained ignorant of what is beyond stars visible from Earth, then came the human brain and with it, the Hubble space telescope.

Rational individuals with access to relevant information now find it easy to accept that the Earth is far older than 6600 years, that it does not occupy a central role in creation, and that humankind is not created in the image of a God, who places more importance on the Earth and humanity than any other beings. The Earth is 4.6 billion years old, Homo sapiens have been on the planet for about 300,000 years and for much of that time, down the food chain, as likely to be eaten, as eating flesh.

This book aims to connect significant events that have made human life possible. I hope this brief overview encourages those interested to explore its research materials and learn from exceptional writers and commentators in science. I strive to share the story that everyone should know, all within 380 pages, or so.

FOREWORD

My writings will reflect the insights of some highly intelligent and well-informed individuals, but my interpretations may not always be completely precise, perspective being one of the differences. However, the theme and the thrust of this book is being mainly guided by their minds, my research and conclusions.

In the beginning, there was... well we can't be sure if it was the beginning. Science, through the application of mathematics and physics, reveals that a massive event known as the Big Bang occurred nearly 13.8 billion years ago, resulting in the birth of our Universe. Theories exist as to what if anything existed before that.

An amazing series of TV programmes, called 'How the Universe Works', goes into detail regarding what is likely to have occurred to produce such energy and defy the laws of physics by using force capable of propelling energy, and therefore potential matter, across the new Universe, faster than the speed of light. How the Universe Works consists of 11 series with a total of 96 episodes. This includes commentary and expert dialogue from eminent scientists.

Life on Earth began to truly appreciate the Universe, when human brains, and the remarkable neo-cortex, achieved enough functionality to explore it. This evolved slowly from Greeks to Copernicus, Galileo to Einstein and Edwin Hubble. It wasn't until the construction, launch and repair of the Hubble telescope that humanity could look back in time, to within a few hundred millions years of the creation. This has contributed to a far deeper understanding of the Universe than ever before, including indicators of age and scale.

In just over 30 years, the perspective of a few has changed because of new knowledge; the shame is, all too few. Billions of humans choose not to make themselves aware or ignore this new knowledge, yet continue to cling to myths about how the Universe was created.

Regrettably, the majority of human minds are unable to fully appreciate these recent breakthroughs. A lack of access, insufficient capacity or education to understand, or a general disinterest, can all be factors. The most worrying issue lies in their inability to embrace new knowledge, choosing instead to resist information that might challenge the beliefs they have held for decades. A modern tragedy in my view.

The work of academic giants such as Galileo, Einstein, Hubble, Newton, Darwin, and numerous other historical geniuses greatly enriches and supports science and its discoveries. Amazing discoveries such as gravity and its effect on time, the effect of mass on gravity, natural selection and its shaping of evolution and who and what we are today. The extinctions of life on Earth have created opportunities for the emergence of current life forms, as a series of disasters contributed to the existence of humans.

As my life is mostly over, I got thinking, what is it in life that has made me what I have become, and more so what are the external forces which have shaped my thoughts and therefore who I am.

I entered the world just like every other human, driven by instincts encoded in DNA, possessing a brain that was a blank canvas for learning, yet brimming with instincts. My parents, brother and sisters were where it started, my wanting to emulate and learn from them, as my instincts taught me. By the age of two, David Eagleman, author of 'The Brain' assures me that my brain would have had more neurons than at any time in my life. More than a hundred trillion. Infants possess an incredible ability to learn.

I had every reason to trust the information I received, whether it was about Santa, God, or learning how to use the toilet; I absorbed it all with enthusiasm, supported and absorbed by my well-nourished brain. I experienced a childhood that I would label as semi-catholic. I received my christening, and all four of my siblings underwent confirmation. As the youngest sibling, I could have been confirmed by ten years old, but I decided against it. I believe my father already suspected that the advantages of religion do not outweigh its drawbacks, favouring a more secular approach to life.

Religion influences almost every individual's life from birth, as it has established the foundation for numerous laws. I was fortunate not to have fallen in love with anyone from the royal family; otherwise, a 1689 law would have prevented our marriage.

I may be trivialising a serious matter that deeply concerns me, but my biggest regret is being born in a time when religion continues to wield significant influence over humanity. I will not travel to any country that bases its modern laws on religion. I believe that law should focus on safeguarding protection, fairness, and morality in society, rather than suppressing diverse opinions.

My time at school did not result in significant achievements because I perceived learning as both complex and monotonous, and I lacked maturity. As I grew older, my fascination with learning deepened, particularly in the field of history and space.

Roman history, the Plantagenet dynasty, and World War II, among others. My exploration of history revealed two key insights: firstly, victors document events and often distort them to boost the egos of their leaders; secondly, historical conflict typically revolves around territory or faith, while notions of right and wrong rarely emerge, except in interpretation, shaped by those who triumphed. If Hitler had won World War II, we may now believe that exterminating a race of people was a justifiable outcome.

The victors usually interpreted the victory as a sign from God affirming that their cause was just. Throughout history, when Christians killed Muslims, many claimed that God supported their actions, and similar claims were made in reverse as well. Truly absurd conclusions.

FOREWORD

'Knowledge has to be improved, challenged and increased constantly, or it vanishes' Peter Drucker

Twenty years ago, we could not know, what we now know. Using evidence which is irresistible to all but the stubbornly dogmatic, evidence which can estimated in numbers and range, but it is also undeniable, we can now give ourselves a fuller picture. Those unprepared to open their minds to the probability that their forebears, and therefore themselves, may be wrong, read no further. There may not be a clear 'right', but numerous definitive wrongs exist that many people continue to hold onto.

*

References, sources and recommended reading, Foreword

University of Warwick - the Astronomy and Astrophysics group - observations of the life and death of similar sized stars to our Sun

Olsen and Straub - American Physiological Society Journal - observations regarding anaerobic life on a young Earth

How the Universe Works - Pioneer Productions / Discovery Channel series 1, edition 5 (1.5) and 9.10

Nasa - The Great Dying (the extinction of 251 million years ago)

American Museum of Natural history - the Earth's major extinctions

NOVA - The Incredible Universe Revealed - discusses the size and age of the Universe

David Eagleman - the Brain, how the brain differs from **toddlers to adults**

Characteristics of state religion - Worlddata.info

Bill of Rights 1689 - English Parliament

Garden of Eden - Britannica

FOREWORD

What is Big Bang Theory – *Space.com*

Epigenetics and the evolution of instincts – *Gene E Robinson & Andrew B Barron, Science.org*

Is it true that Jupiter protects Earth – *Deborah Byrd, EarthSky*

CHAPTER 1

Evolution

—

What is evolution? It is a journey of change without a destination, a journey of survival of species through adaption and change, to changing environments. Humankind is not the finished product, just a progression, on a continuing journey of change until life, human life, and all that follows, can no longer exist on our planet. If Homo sapiens became extinct tomorrow, evolution would continue and without the imposed extinctions of species that humans have brought about. Homo sapiens didn't exist for the first 99.992% of the time life on Earth has been present – we would not be missed.

Joseph LeDoux, an author and scientist, has significantly helped me understand the evolution of life. His book 'The Deep History of Ourselves' offers a fascinating journey through the 3.8 billion years of life on Earth and before. I recommend this outstanding work to anyone who wishes to explore further beyond this publication. It is highly recommended reading. It led me to significant, further research which helped form a further dimension of my perspective.

FORMATION OF EARTH

Scientists believe Earth was formed through a collection of space rock, gases and dust about 4.6 billion years ago, debris and dust brought together by gravity over hundreds of millions of years. In modern times, a time of much less activity within our solar system, between 30-180 tonnes of space debris is added to our planet every day. This is a tiny amount to that added during its formation, which confirms the process even in a much more stable and predictable solar system.

The Earth is approximately $6x10^{24}$ tonnes, or 6 septillion tonnes in mass; to form a planet that large over 200 million years would require an average of around 82 trillion tonnes of mass to be added per contemporary Earth day. This activity gives an indication of how violent and volatile our solar system was at that time.

The large gas planets, including Jupiter, were much closer to the Sun during this time, and as they moved away, they left behind significant amounts of debris. Additionally, previous planets might have existed before Earth, only to be destroyed in collisions with various celestial bodies. Some of the resultant debris, which also contained water, was brought together to eventually form Earth.

For over half a billion years, the extreme temperatures of the surface eliminated any possibility for life to survive and persist, until the Earth finally cooled down. Evidence of life on Earth exists from 4.1 billion years ago; it appears to have struggled to establish itself. Over the course of 300 million years, life fought to survive in a hostile environment; it is believed that life existed many times before it was able to get a foothold on Earth. Life likely emerged and vanished countless times, possibly millions of occurrences over that time. No food supply as we know it existed and first life probably metabolised hydrogen sulphide and carbon monoxide, plentiful on a young, hot, steamy planet.

I further recommend the TV documentary series 'How the Universe Works' for anyone seeking more insight into the dynamics of our Universe, Galaxy, Solar System, and planets. It demonstrates that the existence of life on Earth truly has odds of millions to one, or even many millions to one, possibly billions to one.

Humans occupy a minuscule spot within a fleeting timeframe of a star's formation and its planetary evolution. The 4.6 billion years since our Earth formed is roughly 30% of elapsed time since the Universe formed and a tiny fraction of the expected time until the Universe eventually dies. We inhabit one tiny planet among possibly many septillions (1×10^{24}) of planets and moons that have existed, exist now, or will exist in the future throughout the Universe.

CHAPTER 1

Planet Earth features a lush surface, life-sustaining seas, and a life-supporting atmosphere, representing a rare phenomenon in our galaxy based on our current observations. When we consider the size and scale of the universe and likely length of time it has existed, and length of time it is likely to exist, our story of complex life or similar, elsewhere in the Universe, appears a certainty.

The incredible variety of stars currently present, most with more planets around them, accounts for merely a fraction of all those that have likely existed or will exist in the future.

These numbers indicate that it's nearly impossible for Earth-like planets not to exist in other locations. I express next to impossible for the sake of precision, but in my view, it is genuinely impossible. You will read later in this book about incredible science-based estimates of existing intelligent life in the Universe. As a teenager I recall hearing that having an infinite number of chimpanzees equipped with typewriters, pressing random keys with limitless time, this would ultimately lead to the complete works of William Shakespeare.

We don't need that level of trial and error to explain the existence of Earth and life, but long odds certainly played their part. The development of life and the emergence of complex organisms required numerous factors to align over extensive periods, leading us to our current state.

Setbacks have been numerous, as preserving life on Earth has posed challenges for billions of years. Specific conditions needed to exist on our planet for unicellular life to emerge, thrive, and eventually flourish. Two billion years after unicellular life established itself on Earth, conditions evolved to become more advanced and specialised in order to support the survival of multicellular life, followed by highly complex organisms.

The Sun had to be an approximate size based on our planet's distance from it; if it had been larger, it could have exhausted its fuel before complex life had the chance to evolve on Earth, since larger stars typically consume their fuel at a faster rate than smaller ones. For billions of years, Jupiter's enormous gravity has shielded our planet from many cosmic threats; without it, it is unlikely Earth would have become a life-sustaining world and may have resembled Mars or Venus today.

Around 4.6 billion years ago, not long after Earth formed, scientists believe our planet experienced a collision with a large body, probably a small forming planet; gravity destined the resulting debris to amalgamate into the Moon.

The gravitational pull of the Moon causes the tides in our oceans, which serve as the cradle of life on Earth. Life on Earth might not have existed without the Moon, and it likely would have remained limited to, at best, single-celled organisms. The collision played a vital role in determining the future of life on Earth; without

CHAPTER 1

it, the existence of multicellular organisms seems highly unlikely.

Had the Earth been smaller, perhaps comparable to Mars, it might have met a similar end. Ancient dried-up oceans, seas, and riverbeds on Mars remain clearly visible today. In the past, the planet exhibited a surface dominated by water, much like modern Earth. The signs of oceans over one mile deep have been detected in its northern hemisphere. However, being significantly smaller than Earth, its core cooled, which destroyed its magnetic field, and with it, Mars' protection from solar blasts, much stronger in the younger Sun of three billion years ago.

The Earth's hot core generates the magnetic field that safeguards our planet from the damaging impact of solar winds. The magnetic field emanates from both poles and encircles the Earth, allowing it to sustain our atmosphere and shield it from solar winds and radiation bursts. Without this protection, when this force weakened, it is believed the water on Mars was evaporated, hydrogen and oxygen atoms split apart, into space, when its atmosphere was destroyed by the Sun's radiation.

In terms of size, Venus resembles Earth closely; however, it's regarded as a 'dead' planet when compared to Earth.

Some scientists believe that Venus may also have had oceans, but verifying this is difficult because our knowledge of the planet remains limited. If Venus had oceans, its orbit would come too close to the Sun, making it impossible to retain them. Life on Earth has indeed been fortunate not to suffer the same fate as any life which may have existed or would have existed on either of our nearest neighbours.

Our Sun is relatively stable compared to a substantial proportion of other stars in our galaxy, yet it is still active and aggressive, responsible for the lifeless rocky planets of Mercury and Venus, plus the apparently lifeless planet of Mars.

Having had many circumstances fall favourably for the future development of life, Earth still had some way to go 4.6 billion years ago. Life needed water, but no surface water existed as the Earth was a very hot and hostile environment.

Around this time vast clouds of dust and rock were present in our solar system, left over from the formation of the planets. Up to 20% of this rock is made of water. To touch, rock can feel dry; however, scientists have added heat to powdered rock and measured significant condensed water resultant. By adding heat to rock, water can be released in the form of steam. Earth was made from rock exposed to extreme heat; water in the form of steam would have been resultant.

Massive quantities of water could have reached our planet in one of two ways: it either formed along with

CHAPTER 1

the planet, or arrived here through asteroids and comets. It's clear that a combination of those methods was used; I will let scientists debate the proportions.

In 2011, scientists observed a star forming and noted its size closely resembling that of our Sun. This observation confirmed the expulsion of vast amounts of water vapour carried hundreds of millions of miles from the developing star. When steam travels away from the warmth of the star, it condenses and adheres to space dust and debris. This material contributes to the formation of rocky planets like Earth, Venus, Mars, and Mercury, which coalesce as gravity compresses smaller rocks over hundreds of millions of years.

Numerous theories about the origin of life offer explanations for how the early physical chemistry of Earth evolved into biochemistry. How did life forms develop from inert elements?

It all started with the 'Big Bang' and the hydrogen produced, which formed all atom-based elements, from which Earth inherited its physical makeup. Many of Earth's known chemical elements, which are found in the periodic table, originate as traces of stardust from the Big Bang and formed within stars during the first nine billion years of the Universe, before the Earth came into existence. Our Solar System is essentially 'second-hand', as all the chemicals and elements that make up our bodies, our planet, originated in long-dead stars.

Scientists believe our solar system is made from at least two supernova, a larger collapsing star and a white dwarf which consumed another star.

SINGLE PLANET, SINGLE CELL

In its early stages, the Earth existed as a hot, molten mass. Around 4.2 billion years ago, the surface cooled sufficiently to create a solid crust encased in a primordial atmosphere consisting of carbon dioxide, water vapour, and nitrogen. Oceans originated from steam produced by volcanic activity and water from impacting meteorites. At this time, it is believed life did not exist, but essential elements such as water, carbon sources, and adequate heat for chemical reactions were present.

Water served as the medium for dissolving chemical compounds, allowing them to form new compounds. For example, if you pour table salt into a tumbler of water it dissolves easily – hydrogen and oxygen molecules from water attract sodium and chloride, pulling the salt molecules apart.

Carbon, being a small atom, easily interacts with and forms chemical bonds with other atoms to create compounds that form the basis of living matter; these compounds are generally stable and do not readily dissolve in water. An increase in the surrounding temperature makes it easier for them to break down.

CHAPTER 1

Heat was a readily available resource on a planet cooling from a previously molten state.

These sources of heat were readily available in the form of sunlight, volcanoes, underground magma in the Earth's mantle and bolts of lightning impacting on and heating water. The resultant heat released carbon, which became free to recombine with other atoms. This formed Darwin's primordial soup theory.

LeDoux explains that the replication-first theory argues that self-replicating, prebiotic molecules created the foundations, and resulted in biological replication and metabolism. A molecule only needs to create new copies more quickly than the old copies break down in order to support replication.

It is believed that replicating carbohydrate polymers (complex sugars) were able to do so. The Earth's chemistry at that time likely couldn't support these reactions, but chemistry from space might have been able to. It is possible that 'extraterrestrial sugars' played a role in the emergence of biological replication by becoming the building blocks for nucleic acid, which is essential to the formation of DNA. The essential basis for our understanding of life. Life migrating from extra-terrestrial sources is a Sci-Fi staple; however, more likely is that chemicals from outside our planet mixed with those present on our planet to start the story of life on Earth.

The Earth underwent constant changes after its formation. It has been going through a cooling and change process for 4.6 billion years, born and sustained in a cauldron-like state for hundreds of millions of years. In its early stages, life faced a continuously hostile environment. Single-celled organisms managed to survive and eventually thrive in environments that would likely have been uninhabitable for multicellular life.

*

CHAPTER 2

The Ages of Earth

—

The Pre-Cambrian era spans from the beginning of Earth until 542 million years ago. Spanning more than four billion years, this era included 800 million years during which the Earth did not host any sustainable life forms. Single-celled organisms thrived for more than two billion years before Earth's atmosphere and environment became suitable for multicellular life.

Single-celled organisms with greater efficiency developed improved energy-producing mitochondria earlier, enhancing photosynthesis and boosting their energy supply. From this moment, life's progress on Earth accelerated significantly.

The first game changer, and the reason I'm able to write this book and you are able to read it, came with reproduction. Approximately 3.8 billion years ago, life began the process of replicating itself. It is likely it took sustainable life around 1000 times longer to establish itself through reproduction than Homo sapiens have existed on Earth. The period from 4.1 billion years ago to 3.8 billion years ago likely consisted of non-sustaining life. This was a period of constant trial and error, life and death, over a period of 300 million years, a period of absence of reproduction and transferable DNA.

It is believed that sustainable life as we understand it, started approximately 3.8 billion years ago, relying on cells called LUCA that survived sufficiently to replicate. Around 3.5 billion years ago, LUCA's descendants had split to create what we now recognise as bacteria and archaea. Archaea quickly diverged to create a second kingdom. Both have been in the business of life for a very long time now. Approximately 3.3 to 3.5 million years ago, bacteria acquired the ability to photosynthesize, allowing them to harness sunlight for energy production.

Their ability to thrive in various climatic conditions contributes to incredible success and longevity. Bacteria inhabit every corner of Earth, thriving on land, in oceans, and throughout the atmosphere. These organisms thrive in the moist, warm recesses of our bodies, including environments dense with microbial cells, compared to human cells.

They also exist in snow and ice, as well as in high-temperature waters found within deep-sea vents. Archaea can survive temperatures in excess of 95 degrees Celsius in waters of high salt concentration more than ten times that of normal sea water.

CHAPTER 2

When photosynthesis in single cell creatures commenced more than 3.5 billion years ago it was to later become another game changer for life. Chloroplasts and mitochondria in cells would later produce and store energy with a by-product, waste product, we know as oxygen. The vast majority of animals throughout the history of life on Earth could never have existed without this development.

Tyler Volk points out that cells are always on the border between existing and perishing. They manage to survive by using their metabolism to stay ahead of death. When metabolic waste products are discharged, the result is loss of molecules. To stay ahead of death, cells use metabolism to grow new molecules. If the exchange is at least equal, the cell can exist in its present form.

When a greater number of molecules are created than are lost, the cell experiences growth and increases in mass and size, offering better protection against perishing. A single cell can only expand to a certain size because larger cells demand more nutrients, and they face the problem of a decreased surface area to volume ratio: as a sphere increases in size, its volume grows faster than its surface area.

For a cell this makes it harder for the surface to keep a flow of nutrients high enough to sustain the even larger interior. The unicellular creature divides in half and repeats the process as it approaches its optimum size limit. This maintains a balance between growth and sustainability.

Bacteria and archaea, as with all microbes, reproduce through a straightforward process of cell division. This is asexual reproduction, since only one organism, one cell in this case, is involved. The passing of genes from parent to offspring, this is known as vertical gene transfer. Both sibling cells formed from this division contain the same genes.

For more than two billion years, bacteria and archaea primarily ruled the Earth, which amounts to a period approximately 6600 times longer than the entire duration of Homo sapiens. Then suddenly they had to share the planet with eukaryote, larger single cell creatures that evolved to become multi-cellular.

Bacteria and archaea share a common feature with the early eukaryotes: they all started as single-celled microbes. Yet, several essential characteristics set them apart from their forebears. The main difference being the existence of nucleus – this enabled greater cellular efficiency and energy, which created the foundation for larger, more complex life. Without this development, humans would not exist.

The Darwinian view of evolution highlights divergence – species both new and adapted are created by the cumulative small changes over time, that slowly transform older life forms into new ones.

Plants release oxygen, which allows animals and forms of fungi to thrive, while the carbon dioxide emitted by animals and fungi supports plant life. Tyler Volk calculated that the recycling of carbon dioxide between oxygen-breathing and photosynthetic organisms increased global photosynthesis by two hundred times over what it had been when carbon dioxide was supplied only by volcanoes, rock weathering and other early Earth's chemical reactions.

Later in this book I give an outline regarding the Gaia Principle, developed by James Lovelock and Lynn Margulis, which demonstrates the importance microbes have been to our biosphere and the existence of life on Earth.

MULTICELLULAR EARTH: THE NEXT MAJOR TRANSFORMATION

Single-celled organisms may exist in various locations within our solar system, such as Mars and the moons of gas giants; some are known to contain water in at least one of its three forms. The conditions necessary for single-celled organisms to thrive can be quite extreme, just like they were on Earth three billion years ago. As mentioned earlier, multicellular life needs a much less aggressive environment to survive and thrive. Two billion years ago that environment was starting to take shape.

By producing higher energy output, eukaryotic cells expanded in size, which became an essential factor in the development of multicellular life. The arrival of eukaryotes about two billion years ago changed life on Earth substantially. Their larger size led them to become the first predators, feeding on bacteria and archaea.

A significant rise in oxygen levels was a 'game changer' in Earth's history more than two billion years ago, shortly before eukaryotes emerged due to the increased presence of photosynthetic life-forms that released oxygen, enabling mitochondria to generate more energy. Every complex animal alive today owes its existence to a by-product of the respiration process carried out by single-celled organisms. Our existence is dependent on the waste product of microbes. Animals also survive by using the waste product of plants (oxygen), animal waste product sustains many microbes.

Around 790-810 billion years ago, oxygen levels experienced a second dramatic increase, which was considerably greater than the earlier 'game changing' rise. The oxygenation event during the Neoproterozoic era. This additional growth enabled the evolution of larger multicellular organisms, such as animals, plants, and fungi, and promoted their diversification. Mitochondria function as efficient energy machines in cells, allowing developing life to produce more energy per gene compared to previous life-forms.

Multicellular organisms begin their formation from the zygote. Male and female zygotes conjoin; in humans, each contains 23 chromosomes which ensures the developing fertilised egg contains a complete set of 46 chromosomes. Mitosis describes a type of cell division which resembles the cell division of bacteria and archaea, the most distant ancestors of humans.

Most DNA in a eukaryotic organism is contained in the nucleus, some can be found in the mitochondria. During sexual reproduction in complex life forms, two partners contribute their genes, but the mitochondria primarily come from just one parent, typically the mother.

Both male and female offspring inherit mitochondrial DNA of the egg, but only the female can pass these genes onto her offspring, though some scientists are beginning to challenge this research. Every human female is believed to be linked through mitochondrial DNA to a common ancestor known as Mitochondrial Eve. Scientists suggest that human DNA originates from a shared ancestral group that existed in Africa approximately 140,000 to 200,000 years ago. Various estimates date this phenomenon either earlier or later, but consensus indicates that humans share a common ancestry.

Joseph LeDoux points out that neither Darwin nor August Weismann was aware of the existence of DNA, but Weismann nevertheless was on the right track with his ideas. His theory is, in fact, still widely accepted with certain caveats.

Recent research indicates that drug abuse or stress experienced by a father can increase the likelihood of their offspring becoming vulnerable to addiction and anxiety disorders. This process involves modifying the genes within the father's sperm cells. Natural selection, the altering of DNA at full speed from one generation to the next caused by the introduction of strong external chemical influences.

Factors in the environment that impact genes are commonly known as epigenetic influences, which offer some validation of Lamarck's theory that organisms alter their behaviour based on changes in their environments.

Chemical exposure, including drugs, radiation, and other toxic substances, can disrupt the normal development of an animal's body. The normal unaffected process is that at some point, chemical signals are generated that differentiate cells into the specific types of cells that make up more complex organisms; in humans these make up the body's various tissues and organs (skin, lung, brain, kidney etc). These cells, known as somatic cells, move to their correct places in the organism that is being formed. Biological replication, as we know it today, depends on the coding of the Genome by DNA. Aggressive or perverse conditions can hinder or modify this process.

Among the three groups of multicellular organisms – animals, plants, and fungi – animals are of greatest

CHAPTER 2

interest to humans, maybe because we are animals and have affection for domesticated animals.

Early animals are thought to have emerged in their basic form around 790-810 million years ago. The most widely accepted explanation as to why they existed is the 'colonial flagellate hypothesis'. Ernst Haeckel first suggested the idea in the nineteenth century; however, it was not fully accepted because of insufficient evidence. Contemporary research significantly backs Haeckel's interpretation.

Plants, fungi, and animals all evolved from a common unicellular eukaryotic ancestor. All life on Earth started from unicellular life. Alternatively known as the protists' ancestor of animals, it is an ancient extinct protozoan that is also the ancestor of some present-day protozoa.

Humans have a genetic connection with early single-celled life on Earth 3.8 billion years ago. At the top of the evolutionary scale, we share over 99.5% of our DNA with other humans and 98% of our DNA with chimpanzees. Evidence suggests that our lineage traces back to a shared ancestor from around seven million years ago. Humans also share 8% of our DNA with early viruses.

The evolution from protozoa to multicellular organisms created higher levels of energy in cells, which in turn required more oxygen. The dramatic rise in Earth's concentration of atmospheric oxygen, about 800 million years ago, coincides with early animal life.

OXYGEN, RAPID PROGRESS AND DISASTER

The next major era on Earth was the Palaeozoic Era from 542 million years ago to about 251 million years ago, ending with the worst extinction of any post-Cambrian era. During this period, and prior to disaster, life flourished in the oceans, and complex organisms diversified broadly on Earth.

The extinction of 96% of life in Earth's oceans marked the end of this period, but the Palaeozoic era had offered some protection for life due to the diverse range of species it generated, thereby making life on Earth more resilient to abrupt changes. Evolution used the gene pool of life remaining after this prehistoric Armageddon, to once again create myriad flora, fauna and fungi.

Approximately half a billion years ago, the partnership between fungi and plants allowed for nutrient sharing. Plants supplied fungi with sugars via photosynthesis, while fungi provided plants with minerals extracted from the existing environment. True symbiosis came into existence. Plants had been releasing oxygen on land as a by-product of photosynthesis for about 50 million years before amphibians were able to evolve, to walk on land. This enabled respiration and metabolism to take place on land, another 'game changer' for life on Earth.

CHAPTER 2

LeDoux leads us through the answer to the following question. What process led to protozoan life evolving into the first animals, such as sponges, and subsequently how did sponges give rise to cnidarians like jellyfish, sea anemones, or corals? How did a cnidarian evolve into a bilateral invertebrate? Bilateral invertebrates then evolving into vertebrates. Evolution and natural selection driving life to become more complex and adapt to new environments. Natural selection ensured adaption of living things to their changing environments.

Vertebrates originated in the seas and oceans. Fish are often described as the first vertebrates, and while they certainly represent the oldest living vertebrates, a creature that existed before them bridged the gap between invertebrate chordates and fish. A fossil dating from 530-540 million years ago identified this creature, haikouella.

Approximately 370-380 million years ago, a significant event occurred in the evolution of fish that would later impact on humans. They started to evolve into a new group of vertebrates that adapted to living on land. This now extinct creature represented an evolutionary form between fish and tetrapods, the four-legged animals; their ancestors would later include amphibians, reptiles, mammals, and birds that stand on two legs.

A new classification of animal developed around 320-340 million years ago. These creatures belonged to the group of amniotes, which raise their pre-birth young within an internal amniotic sac filled with fluid, allowing embryonic, foetal development. New types of lungs enabled them to draw oxygen from the air rather than reliance on water. This allowed animals to give birth away from water.

Synapsids are believed to be the first group of reptiles to branch off from amniotes. Mammal-like reptiles existed as the ancestral species, a very significant development for humans as it formed the group from which mammals would later evolve.

The extinction event 251 million years ago (known as great dying) marked the end of the Palaeozoic era and heralded the beginning of the Mesozoic era. The Earth's temperature rose significantly, leading to a mass extinction that predominantly affected marine animals and larger land creatures. Among the species that survived, new opportunities arose, and the most notable terrestrial survivors were cynodonts.

Roughly the size of a large dog, they displayed features and characteristics linked to the modern day mammals they would eventually evolve into, they developed jaws, eye sockets, ears, indicating skeleton structures along with the emergence of hair and mechanisms for internal temperature regulation, typical in mammals.

Cynodonts separated into herbivores, omnivores and carnivores and further spread across various ecosystems and climates worldwide. The varying ecosystems would drive evolution and natural selection to accelerate

CHAPTER 2

diversity, eventually leading to the emergence of true mammals by 200-210 million years ago. Mammals have existed on Earth for around 205 million years out of its total 4.6 billion years, which represents only about 4.5% of the planet's history. The growth of mammals came with an increase in brain size, relative to size.

German anatomist Ludwig Erdinger made a significant observation that influenced perspectives on the evolution of the vertebrate brain in the twentieth century. He observed that the hindbrain and midbrain displayed remarkable similarities among various vertebrates, while the forebrain exhibited differences in size and complexity in a progression from fish to mammals.

Erdinger suggested that this was due to sequential expansion by layering of new structures in the forebrain as reptiles diverged from amphibians, and again as mammals diverged from reptiles, with further changes continuing in mammals, eventually resulting in the proportionately largest, most complex brain, the human brain.

Plants were also advancing. It is estimated that there are 400,000 types of seed-bearing plant on Earth, the first appearing 250 million years ago. Gymnosperms (naked seeds) and flowering plants, known as angiosperms, emerged approximately 130 million years ago.

Insects and flowering plants formed a symbiotic relationship. We are familiar with the honeybee and its collection of nectar, the flowering plant in turn placing pollen on the body parts of the bee which fertilised plants they visited, thereafter. Many species of insect have fulfilled this role during the evolution of plants and insects.

This brief synopsis of life on Earth was not without incredibly damaging setbacks. Over the past 450 million years, life on Earth has survived and rebounded from at least five significant extinction events. These include a number of different causes from extreme volcanic activity, strong acidic environments and the latest, caused by an asteroid. Nevertheless, each of these elements contributed to the extinction of a significant portion of life on Earth. Survival favoured the relatively small and highly adaptable individuals.

The most significant documented extinction is estimated to have happened roughly 251 million years ago, which I referenced earlier. About 96% of all life became extinct. Contrast that with the extinction event that occurred 66 million years ago, leading to the disappearance of 76% of all life. Around 445, 360, and 200 million years ago, three additional disasters caused the extinction of between 75% to 86% of all life forms.

The major extinction we are most familiar with, 66 million years ago, which destroyed the dinosaurs was, therefore, in comparison, much less destructive to life than that of 251 million years ago, but was essential for the eventual rise of mammals then primates and

CHAPTER 2

humankind. The removal of large and voracious reptiles likely paved the way for mammals to evolve and ultimately dominate the planet.

The five known extinctions: Ordovician-Silurian extinction about 445 million years ago; the Devonian extinction 360 million years ago; the Permian-Triassic mass extinction of 251 million years ago; the Triassic mass extinction of 200 million years ago.

Then, 66 million years ago, an asteroid caused the Cretaceous mass extinction. The Yucatan Peninsula, near present-day Mexico, experienced a tremendous impact from an object measuring six to seven miles wide, which travelled at alarming speed and caused the extinction of dinosaurs along with numerous other species. The enormous impact generated dust that obscured the Sun for several years, creating an almost immediate toxic environment for most species on Earth. This marked the conclusion of the Mesozoic era and the beginning of the Cenozoic period, which is when Homo sapiens emerged, existing for less than 0.5% of the start of the Cenozoic period to present day.

Both the human species and all large mammals likely owe their existence to catastrophic events. The last of which, enabled small mammals to thrive, having lost several major reptilian predators. Small mammals evolved through natural selection into the diverse range of mammal species that populate our planet today, including humans.

One can assert with confidence that there were five distinct points in the last half billion years, where extinction events redirected the process of evolution. Without the earlier extinctions dinosaurs may not have existed. If there had only been zero to four extinctions instead of five, humanity as we know it probably wouldn't exist, nor would the numerous deities that humans have conceived.

It is also conceivable that an intelligent lifeform may have evolved in place of humans. Natural selection has played a crucial role in the development of humankind, drawing from the genetic materials available from life on Earth 66 million years ago.

This process has repeated itself after every mass extinction. It will happen in the aftermath of the next extinction. It is unlikely that humankind would survive the resultant devastation of any of the five previously recorded extinctions. Yet life on Earth would continue following another setback soon to be consigned to history, as were the others. If these extinctions had not occurred, nature would have worked with a different diversity of DNA. It is reasonable to assume life would be very different today. Animals possibly as intelligent, or more so, than humans, and likely to appear very different, having evolved from different DNA.

Dolphins have significant brain capacity; however, without eight fingers and two opposable thumbs, plus access to dry land, they have not conquered their environment to the same extent as humans. If dinosaurs had continued to evolve on land for another 66 million years, they would have experienced further divergence, but the degree of that change will remain a mystery.

Had the extinction of 66 million years ago been as extensive as the extinction of 251 million years ago and killed 96% of species rather than 76%, then the additional species extinction would have almost certainly included far more mammal species, which would likely have been the distant ancestors of humankind.

A less serious disaster and our early ancestors may have remained merely a food supply for more, larger and efficient predators, had those predators not been wiped out by cataclysmic incidents. Should the extinction event of 66 million years ago never have occurred, or if it had been less catastrophic, it's possible that humans would not be present today.

The future of Earth and humanity raises concern due to the similarities between the extinction event that occurred 251 million years ago, the most destructive in evidenced history, and the current effects of global warming. One popular theory is that Extinction was caused by volcanic activity in what is now Siberia. For more than 60,000 years, massive quantities of carbon dioxide flooded the Earth's atmosphere. This caused a 'greenhouse' effect of warming the atmosphere to such an extent the seas, the cradle of Earth's life, turned acidic and hostile to life.

Sea hydrogen sulphide from algae, volcanic carbon dioxide and chlorine gas from magma meeting trillions of tons of salt crystals from dried up seas! For ten million years no significant coal reserves were laid down, as no trees could grow; it took many more millions of years for the Earth to cool and recover.

The similarities end with timescales. Earth is warming at an unprecedented rate, likely due to human excesses and emissions, which shortens the timeline for disaster to hundreds of years instead of tens of thousands of years.

Today, a fifth of mammals and a quarter of reptiles are at risk; the World Wildlife Fund highlights an even gloomier forecast. The rate is elevated for amphibians. Animal extinctions currently occur at a rate 12 times higher than the historical average over recorded millennia. The loss of one species has a knock on to others which might prey on them or have other dependencies.

An extinction caused by humans, including humans, would only mean a delay for the progression of Earth and natural selection. As it did 251 million years in the past, the Earth will endure and recover. Humans will soon become a distant memory, having once dominated the Earth for only a brief period compared to the longer reigns of bacteria or even reptiles.

Approximately 17 million years after the great extinction, around 234 million years ago, a period of heavy rainfall lasting two million years transformed the landscape and

CHAPTER 2

led to the emergence of forests, marking the greening of Triassic Earth. The Carnian Pluvial Episode refers to a significant period of increased rainfall and climatic change during the Carnian age.

New species of reptiles and amphibians evolved, leading to 140 million years of dinosaurs, representing more than 450 times the length of time Homo sapiens have existed on Earth. In the time of the early dinosaurs, Earth's surface had a different configuration, featuring a single major continent known as Pangaea instead of the six or seven continents we have today. It also had more oxygen in the atmosphere, needed to support large land-dwelling reptiles.

The Earth's surface is situated atop tectonic plates that can be as expansive as present-day continents. Over 200 million years, enormous forces driven by thermal pressure from deep within the Earth split this landmass apart. Sea level fluctuations caused by freezing and melting events influenced the proportion of land to water, on Earth's surface.

Today, the Indian subcontinent forms part of the Asian continent. But more than one hundred million years ago, that area sat thousands of miles to the south, nearer to southern Africa. The subcontinent, propelled by tectonic and volcanic forces, moved northward at approximately five to 20 centimetres annually, translating to a maximum of a kilometre every 5000 years. This movement caused it to collide with what is now southern Asia between 35 and 55 million years ago, over the course of millions of years.

Massive forces driving the subcontinent kept pressing northwards and the Himalaya mountain range was created through the pushing of landmass, against landmass, forcing land upwards when it could no longer progress forward. The Himalayas were once at the bottom of an ancient ocean but forced upwards to now be the tallest mountain range in the world. Ancient fossils of seagoing creatures can be found near the Himalayan highest peaks – once at the bottom of ancient seas.

The journey to this point was a story of turmoil, disaster, extinction and survival of the fittest. Eventually we arrived at humankind's more recent ancestors. Life managed to cling on through many challenges and assisted by natural selection created sustainable life for the future. But what is life…

*

CHAPTER 2

References, sources and recommended reading or research, Chapters 1 & 2

Observations of the early Earth, Evolution and the Human Brain - *The Deep History of Ourselves* - Joseph LeDoux

How the Universe Works - *Pioneer Productions / the Discovery Channel 4.6 and 10.3*

New Scientist - *observations of evidence of life on Earth 4.1 million years ago*

UCR College of Natural and Agricultural Sciences - *observations on Earth and Venus*

Earthsky.org - *observations on planet Mars*

Robert Irion - *Science Adviser - observations on the rocky planets*

NASA Science - *observations on Mars, Mercury and Venus*

Edinburgh Botanical Gardens - *Angiosperms*

Mass accumulation of Earth from interplanetary dust, meteoroids, asteroids and comets - *Science Direct*

What is a Genome - *Yourgenome.com.org*

The Thermal Limits to Life on Earth – Harvard University

The Effect of Oxygen Concentration on Photosynthesis in Higher Plants – Wiley online library, Olle Björkman

Predatation between Prokaryotes and the Origin of Eukaryotes – Wiley Online, Y Davidof, & E Jurkevich

How Does Environment Affect You DNA – Royal Institution

The Single-celled Ancestors of Animals, A history of hypotheses – Thibaut Brunet & Nicole King, Howard Hughes Medical Institute, Dept of Molecular Biology

A Glimpse into Early Animal History – Science, Martin Enserick

First Fossil Fish that Crawled onto Land Discovered Life by Bob Holmes

Origin of Amniotes and Amniotic Egg – Robert R Reisz & Tea Maho

Carnian Pluvial Episode – An Overview – sciencedirect.com

Early Earth - Lumen Earth Science

CHAPTER 2

UCL Earth Science – Alex Kraus, BJW Mills, University of Leeds – Neoproterozoic observations

National Library of Medicine – Chromosome observations

Genetics Utah Education – Learn Genetics

How the Universe Works 10.8

Evolution and Development - Stanford University

Carnian Pluvial Episode, That Time it Rained for 1-2 million years – Gen Engineer

People are Confused why there are Fossils at the top of Mount Everest – Plate Tectonics, National Geographic – IFL Science

Animal Life, early origins – Imperial.ac.uk

Evidence for Early Life in Earth's oldest Hydrothermal Vent Precipitates – Dodd, Papineau, Little, Nature.com

Explaining dramatic planetwide changes after world's last 'Snowball Earth' event – University of Washington – Hannah Hickey

Diverse forms of life may have evolved earlier than previously thought – UCL.ac.uk

Evolution – ScienceDaily.com

Americanaddictioncenters.org – observations on additions and stress being passed from father to child

US Geological Survey.usgs.gov – observations on Paleozoic era

Continents and Supercontinents – J J W Rogers and M Santosh Oxford Academic/ National Science Review

New Scientist – observations on early life on Earth

Volcanically driven lacustrine ecosystem changes during the Carnian Pluvial Episode (late Triassic) – pnas.org, J lu, P Zhang, J D Corso & J Hilton, Proceedings of the National academy of Sciences

Nature.com, scientific report – India drift

What is Evolution – your genome.org

CHAPTER 3

What is Life?

—

'Living organisms exhibit movement, respiration, sensitivity, growth, reproduction, excretion and nutrition, is what life does but not what it is'?

Paul Nurse

*

We have discussed how life came about, but what is it? Paul Nurse gives an excellent summary in his book 'What is Life'.

Nurse starts by discussing microbial cells. They inhabit every environment, from the high atmosphere to the depths of the Earth's crust. Without them, life would come to a standstill. Waste decomposition, soil building, nutrient recycling, and nitrogen capture from the air support growth in plants and animals.

Humans have about 25-35 trillion human cells and probably more microbial cells (lodgers). These organisms reside within us and influence many aspects of our lives, including our digestion and ability to fight illnesses. We could not survive without their contribution.

Nurse discusses the process of cell division. It serves as the basis for the growth and development of all living organisms. This marks the initial crucial phase in transforming a homogeneous fertilized egg from an animal into a cluster of cells, which then develops into an organised and complex living organism, an embryo. All living organisms evolved from unicellular creatures and all life today emerges from a single cell at the point of conception.

Every gene a lifeform possesses comes in a pair, with one being passed down from each biological parent. At the time of conception, sperm delivered them to the egg, resulting in fusion. Numbers of chromosome pairs vary with complexity of living organism – pea plants contain 14 chromosomes in each cell, an earthworm 36 chromosomes, whereas humans possess 46 chromosomes, 23 from each parent.

At the core of each chromosome in living organisms lies a single molecule of DNA, which is unbroken. These can reach substantial lengths, with each containing hundreds or even thousands of genes linked in succession. Human chromosome contain a string of more than 1300 different genes – if a piece of DNA was stretched out, it would measure more than 8cm in length. All the DNA in a single cell is around two metres in length.

CHAPTER 3

The 46 chromosomes in each tiny cell combine to create more than two metres of DNA. It all fits into a cell that measures no more than a few one thousandths of a millimetre across. Equivalent to packing 40km of thread into a tennis ball. If you could join up and stretch out all the DNA coiled up inside your body's 30 trillion cells into a single slender thread, it would be about 20 billion kilometres long. That is enough to stretch from the Earth to the Sun and back more than 60 times.

We can assert with confidence that our primary role in the evolutionary process is to store vast amounts of information in our cells and pass this information down to the next generation. If you imagine an X and you are the dot in the centre, your ancestors' DNA is contained in the bottom half of the X, your parents' DNA just under you on each leg of the X, grandparents under them and so on.

Above you on the upper legs are future generations, children, grandchildren, plus the 'yet to exists'. You are one small fulcrum in a mass of humans, a pivotal point for the passing on of your genes but also those of your parents and Ancestors, keeping their information alive. The bottom of your X stretches back 3.8 billion years in an unbroken chain of evolution.

It appears we are a vehicle, a cog in something much bigger and we are disposable as all life which has fulfilled its purpose, is nothing more than a chrysalis shell or a shed snakeskin, necessary for a time, yet fully disposable.

DNA is what nature values and strives to preserves, individuals appear meaningless on death, a commodity once with value, which was lost.

For those who die prematurely or for whatever reason do not have offspring, stand at the top of a point which never continues onwards into an upper v of the X. They break an unbroken chain of billions of years. This hardly matters to the evolutionary process as the information will get passed through other and many X scenarios. It is the species, not the individual, that is of key importance to natural selection.

Fifty percent of parent genes passed to a child, an average of 25% into grandchildren, until our great, great, great-grandchildren when our genes have approximately a 3% input. But the vast majority of animals don't reach maturity and therefore don't reproduce.

If you are reading this, you have been born a human and reached an age and educational level which enables you to learn complex issues and articulate them. Viewed from a human perspective, that is a lottery win in terms of life, when you consider the abundance of different life forms, all with consciousness and realised lives.

If you live in a westernised country and you take your health seriously, you may have in excess of 80 years of consciousness, over 29,000 days. Significantly longer than most animals on Earth. Most of us don't wake up in the morning wondering how will we eat that day; it's

CHAPTER 3

more about choosing what we will eat and how we will spend the surplus 23 hours. Humans have developed significant differences from all other species, yet they still share survival instincts through their common DNA.

Life for all is short, compared with Universe time elapsed. Many insects live for a matter of hours on reaching sexual maturity, their whole lives reaching a crescendo when genes are passed on and the life objectives set by natural selection and evolution, are achieved.

Evolution and brain capacity growth have allowed humans to create enriched lives and form deeper, more meaningful relationships, that go beyond mere reproduction. We possess the ability to communicate, build friendships, make plans for our futures and many other skills. We have numerous opportunities for experiences, and variety in life is plentiful. Nevertheless, humans have the same task set for them by evolution: survival and continuation of the species.

Evolution provides improvement modifications for existing and future life out of this process, through adaption and successful natural selection. A giraffe did not always have a long neck – it happened over thousands of generations reaching higher and higher into trees for food to gain advantage over other large mammal leaf eaters. Gradually, necessary behaviour for the purposes of survival changed the DNA of the giraffe, ensuring that with each generation necks got a fraction longer, ensuring the species was able to compete and survive.

Camouflage and the ability to change colour to adapt to environment, running fast to catch prey and just about every feature of animals, plants and fungi alive today was achieved by developing or changing messages carried by those four precious chemicals: Adenine, Thymine, Guanine, and Cytosine, the chemical bases for the coding of DNA.

Is there a chance of our DNA being exactly replicated in the future, resulting in a form of reincarnation? DNA can be replicated through cloning and identical twins also share it. However, twins and clones are individuals and have their own 'realisations'. The evidence suggests that even if someone perfectly replicates your DNA, the duplicate would not be you. The 'who we are' is discussed in other parts of this book. We are shaped by experience, learning even by the books we read and movies we watch. An injury or growth in our brain, drugs, alcohol etc can change us significantly.

DNA encodes the information that cells and entire organisms require to grow, sustain, and reproduce. Secondly, it needs to generate copies of itself with precision and dependability, allowing each new cell and organism to receive a full set of genetic guidelines.

Nurse makes the point that DNA has a helical structure, which you can think of as a twisted ladder, and this explains both of these critical functions. Each rung of 'the ladder' consists of links that connect pairs of chemical molecules known as nucleotide bases. These

CHAPTER 3

bases come in four different types as mentioned above, Adenine, Thymine, Guanine and Cytosine, abbreviated, A, T, G and C.

The order in which these four base chemicals appear along each of the two rails, or strands, of DNA ladder, function as an information-containing code. The message might be an instruction to produce a pigment that will determine hair colour, make the petal cells of a rose, red, or make a plague bacterium more virulent, for example. The cell obtains these messages from DNA by 'reading' this genetic code and putting that information to work.

Natural selection, the evolutionary process which has seen our world arrive in the present with myriad millions of species with billions of variations has relied on the adaptability of Cells, Chromosomes, DNA to harmonise with changing environments. For example, as the world became colder in ice ages, those mammals best equipped to keep themselves warm through changing bodily processes, for example growing thicker fur, and becoming more efficient in collecting sufficient calories of food, would be more likely to survive and produce future generations.

Natural selection through its DNA alteration processes can appear to respond slowly in changing living things to adapt to their environment; this will have been the cause of some species to become extinct. Environmental changes may have occurred too rapidly.

Dogs showcase how quickly humans can alter DNA through crossbreeding for various purposes over just a few generations. It was mentioned earlier that children might inherit a predisposition for drug dependence from their fathers due to changes in DNA that occur within a single generation.

As a young man, I recall watching a television programme about a tribe in Papua New Guinea that had maintained no contact with the outside world for hundreds of generations. The population remained hunter-gatherers because of the abundant food supply and the natural limitations imposed by terrain and environment, which did not create a need or opportunity for farming, despite the presence of some domesticated animals. Indigenous people began to show signs of tooth decay within three weeks of westernised humans living among them. It was clear that little resistance existed to the bacteria that rot teeth.

Tooth decay commonly affects both developed and underdeveloped countries, but some level of resistance can be observed. Even without dental care or fluoride in the water, teeth in Western societies may take several months or even years to decay when exposed to processed foods, sugars, and bacteria that create tooth decaying acids as they feed on waste food binding to them.

The case of the New Guinea tribe experiencing rapid tooth decay illustrates how resistance to such diseases has developed in populated areas away from isolated populations, aided by genetic information passed down through DNA. The lack of bacteria responsible for tooth

CHAPTER 3

decay in the jungles of New Guinea eliminated the need to develop any level of immunity. Although not tested at that time, it is likely the Indigenous peoples had a greater tolerance of spider and snake venom etc having been exposed for many generations, than Western cultures where few venomous interactions take place.

Many humans will be familiar with fungal infections, the most common being Candida. Various types of fungi commonly inhabit plants and animals, with some ant species supplying food to the fungi and feeding exclusively on fungi. This illustrates how change has occurred over hundreds of millions of years. Fungi have existed for more than a billion years before the emergence of animals, they then adapted their strategies to use animals as both a food source and habitat, whilst supporting the lifecycle of some animals.

Fungi and animals now adapt to new environments through genetic programming that occurs before birth. This process assists in the development of new species; at some point in the past, fungi split into those which thrived in one environment and those who thrived in others. A similar theme to the human and chimpanzee ancestral split around seven million years ago: one branch remained living in trees, another walking upright on the savannah. Splits in species being a permanent and continuing feature of evolution and natural selection.

Cells are wonderful lifegiving things – not only do they contain the building blocks of life, they produce the energy needed for life to function. Most of us will be aware of the main food groups, fats, proteins, carbohydrates / sugars etc but few will be aware of how cells produce energy food stuffs. Without getting too complicated, the cell has a process called the Krebs or Citric Acid cycle where cells convert sugars including glucose, into energy.

A rapid, repetitive, circular chemical reaction occurs when various chemicals convert into new elements by typically adding different molecules. Flavin adenine dinucleotide, nicotinamide adenine dinucleotide, water, carbon dioxide, vitamins and other molecules, which is getting a 'bit heavy' in complexity for the ambitions of this book. But in summary, this initiates an oxidation process inside the cell that generates energy. This process is in place in every cell when energy is needed for all kinds of functions, through metabolism and respiration.

DNA carries a type of genetic 'memory' that is transferred from one generation to the next. Not in the sense of memories of holidays and good times with friends, but learning data necessary for survival passed from both parents to offspring.

INSTINCTS AND NATURAL SELECTION

Darwin gives instances of bees building honeycombs as an example of instinct in his book On the Origin of

CHAPTER 3

Species; they also dance to demonstrate the position of food sources that are so accurate and detailed they can direct a fellow bee hundreds of metres or even kilometres, to a specific plant with available food.

Other examples include various insects and reptiles that emerge independent and solitary from the moment they are born. Many spiders are independent from the moment they hatch, inheriting instincts from their DNA that evolved over millions of generations. This adaptation allows them to construct webs to catch prey, kill it swiftly, seek protection from predators, and reproduce cautiously to ensure they don't become a meal, themselves.

Numerous predators avoid brightly coloured prey, as it frequently indicates danger in nature. Certain species of South American tree frog possess potent toxins that can be lethal to large mammals upon skin-to-skin contact.

Non-toxic animals will become brightly coloured to mimic the protection afforded to brightly coloured venomous animals. A capability that evolved through DNA over numerous generations. A key instrument of survival passed through chemical code.

The biologist Karl Niklas argues that in the process of creating new species, natural selection "acts on functional traits" rather than on the mechanisms that generate them. In other words, 'conserved genes' translate into general functions that keep an organism alive.

I believe, this could be considered 'unconscious intelligence' strategy and tactics. Survival strategies reflect the basic requirements of cellular life and common in all living organisms. Because different organisms have different body structures that reflect ways their species has been tailored to environments through natural selection, they possess different behaviours or tactics to implement a survival strategy. All to varying extents, pre-programmed before birth.

Organisms move toward environments that promote survival while avoiding harmful elements that would likely create threats or disadvantage, the selective behaviours passed on through beneficial genes, to their offspring. What they pass on is specific biological characteristics related to survival.

The nervous system in animals informed decisions of approach or withdrawal beyond toxic responses in the case of unicellular organisms. Neural control enabled a more precise decision regarding approach and withdrawal methods to meet specific survival needs. As animals developed more complex and responsive nervous systems, their behavioural options also became more advanced.

A male hippopotamus in breeding season will know when another male is significantly larger and withdraw from danger. When more evenly matched, the hippo instincts will drive the hippo to 'make a decision' to engage in a fight that often ends in death, with the balance being

CHAPTER 3

the possibility of fathering the next generation. This is a calculated risk with two potential outcomes: one provides the chance to breed, while the other probably results in becoming food for scavengers. As reproduction is the hippo's main purpose in living, the risk must be taken when he is somewhere near his prime, so meeting a larger hippo at that time is unfortunate.

When organisms satisfy specific survival needs related to energy, nutrition, fluid balance and defence or when they engage in sexual reproduction, they are, in the most basic sense, typically using neural circuits to call upon specific approach and withdrawal behaviours that are matched to their survival needs by their specific genes.

Adult humans and in many cases humans as young as toddlers instinctively know that a spider or snake are species to be wary of and avoid. Apes, our closest ancestors, also know the animals to avoid from an early age. This primeval fear has been passed and instilled in our genes for tens of thousands of generations and more. As we grow and learn, animal instincts mix with experience to improve response to the environment and heighten chances for survival.

A team of researchers from the Max Planck Institute in Germany, performed a study with 48 six-months-old babies. When shown photographs of spiders and snakes, on a white background, the babies' pupils consistently dilated more than when shown photographs of fish and flowers. Dilated pupils are associated with noradrenergic system in the brain that processes stress. A clear indicator that at six months, babies were more intimidated by spiders and snakes than fish or flowers.

The human brain can develop irrational fears through experience and because of the size of the human brain and its capacity for developing thought processes, this can cause disadvantage. My wife is one of only two people I know who has a fear of birds, especially in flight and in close proximity. When she was a child, her cousin grabbed a flapping budgie by its feet and pressed the frightened bird against her face. An 'in life' learning experience which has been promoted to primeval fear, in her mind, with accompanying reflex actions, on the same level as others may react to venomous insects, spiders or snakes.

Over billions of years, our dynamic planet has changed environments, necessitating life to adapt; the survival of species hinged on these adaptations. Natural selection and DNA were 'bedfellows' in that change process. The Arctic fox features a warmer coat than the red fox; over generations, its DNA has transmitted information that encourages the development of thicker fur for survival.

Fortunately for the Arctic fox, ice ages progressed slowly so the transition would have had a comfortable timescale. Other species have not been so fortunate. It is believed that some species of flightless birds in the Pacific islands were made extinct when humans unwittingly introduced black rats to the islands. No time was available for the

CHAPTER 3

birds to adapt from ground nesting to more inaccessible sites. The theft of easily accessible eggs led to extinction as the birds could offer little defence, especially when absent from the nest in search of food.

DNA genetic messages have 'engineering' qualities when an animal embryo is developing in the egg or the womb, genetic codes will inform where organs are placed (somatic cells), what blood vessels go where, and, to what extent development takes place pre-birth. Preparing some animals for immediate self-sufficiency or preparing others for survival and post-birth growth first, self-sufficiency later, as in the case with many mammals.

Many millions of species, including animals, plants, and fungi, have likely gone extinct in the 1.5+ billion years since multicellular life emerged on Earth. The effect of this had been to give the species that survived more chance of survival and development through natural selection as less competition for food and other survival essentials existed.

THE VOLUME OF SURVIVAL

Animals, especially insects, fish, amphibians and reptiles, generally produce many more offspring than mammals or birds. Plants and fungi can release thousands of seeds or spores in an effort to guarantee that at least one new offspring survives to adulthood. Evolution has engineered this, over millions of years, to enable survival of species. In areas with high attrition and mortality rates, natural selection has elevated birth rates to enhance the likelihood of producing sufficient adults who reach sexual maturity and can generate future generations.

In many species, young have less than a one per cent chance of survival to maturity, demonstrating how expendable individuals are to the evolutionary process. The aim, having been to arrive at sexual maturity, for a small percentage, enough of a species to ensure continued levels of future generations, with little regard for the statistical casualties. Nature is sending a clear message, that the individual counts for little against the survival of a species.

If sacrificing hundreds of immature offspring allows one or two sexually mature adults to thrive, then it must be done. Nature strives for equilibrium; while underpopulation risks extinction, overpopulation jeopardises resources that should support future generations.

Species of ants, bees and other insects produce few members with the ability to reproduce; the vast majority of others are providers of accommodation, food, protection and raising of young. Natural selection broadened the 'survival by volume' equation to raise the chances of a species surviving and thriving. The queen carries millions of male zygotes from her first sexual encounter and is able to produce hundreds of thousands of fertilised eggs

CHAPTER 3

over her lifetime, many thousands per day in some cases, growing the colony and replacing those lost in pursuit of the tasks programmed by their DNA.

The vast majority have no opportunity of enabling their own genes to participate in future generations, yet will sacrifice their lives for the offspring of their queen. What a very human type of behaviour, except that it is pre-determined before birth, whereas humans, on many occasions, make a conscious and considered decision to give their lives in wars, live celibate lives, or to live without having their own children, yet support the children of others – second marriages for example.

Humankind has suffered from high mortality rates, relative to contemporary humans, especially in infants. People in Western societies with access to healthcare, nutrition, and welfare might find it challenging to envision the significantly poorer living conditions that existed only eight to ten generations ago.

Victorian Britain experienced high mortality rates, particularly in cities and amongst children, due to a minimal welfare state, inadequate nutrition, poor sanitation, and widespread contagious diseases, unlike modern Britain. Similar to the places on the Indian subcontinent today, the 'underclasses' of Britain relied on their children to support them when they could no longer support themselves.

This led to a vicious circle – having more children than they could afford to support, in order to enable some to survive to maturity, but the more they had the more they were likely to lose prematurely.

In modern Britain the vast majority of children, at birth, can reasonably expect to outlive their parents, but this was not the case just 150-200 years ago. Infant mortality was more than 200 per thousand in early Victorian Britain. In 1830, seven years before the start of the Victorian era, 57 children out of every 100 were dead before the age of five. When added to post-infant mortality, prior to reaching adulthood, this describes a desperate situation for those living in poverty.

Contemporary humans have a raised chance of embryos reaching birth and then maturity to adulthood – developed civilisations and advances in medicine, nutrition and hygiene levels have greatly assisted this. Natural post-copulation and pre-conception have been less influenced by humans; however, fertility treatments have had a great influence in enabling parenthood.

Human males produce an average of between 40-150 million sperm per ejaculation aimed at fertilising one ovum. An average of about 8% will reach the fallopian tube from the vagina, as the woman's immune system will attack and kill most of the sperm. It is a hazardous journey which has no guarantee of success.

Natural selection has prepared human DNA to trigger the production of incredible numbers of sperm to provide a realistic chance of successful fertilisation. Hundreds

CHAPTER 3

of millions of sperm, each carrying half the required information to produce a human, sacrificed on the 'altar of volume' for the sake of survival of the species. The individual sperm is meaningless compared to the single sperm which proves to be successful.

Animal reproduction evolved over billions of years, while humans have introduced the concept of planning into this process in recent times. Male births have traditionally gained strong preference to female births in both the Indian subcontinent and China, regions where the poor face scarce resources and limited nutrition. Having adult males support their parents in old age is considered much more beneficial than having females to do so. This has caused extreme behaviour in some, resulting in the aborting, killing or giving away female offspring.

Chinese authorities enforced a one-child policy for many years, resulting in increased abandonment of female infants or abortion of female embryos, which led to a significant shortage of adult women to form relationships with adult men. And it was a further contribution to the Chinese government's main goal of lowering birth rates and subsequently reducing the population, as many males lacked the opportunity to become parents.

Species similar to humans, which typically give birth to a single offspring per pregnancy – sometimes two and rarely more – have developed more effective survival strategies. If we had not made these advancements, our species might have confronted extinction from inadequate birth rates, unable to sustain the population, and compensate for early deaths.

Primates, such as humans, elephants, and various other mammals invest significant amounts of time raising their offspring to adulthood. Many primates and elephants are endangered because having previously cracked the survival code, conditions and environments have mitigated against them, whilst humankind has flourished. The two developments clearly relate to each other.

Humans discovered larger communities, developed civilisation, practised farming, installed sewerage systems, improved healthcare and enhanced nutrition, all of which significantly increased survival rates, particularly in the last 200 years. Elephants, primates and many other animals found an expanding humankind.

One outcome of the above was a rise in human life expectancy, along with an increase in fertility years during that lifespan. Individuals now have higher chances of survival, and they can reproduce for a longer duration.

The modern record of births produced by one female human is 44 babies over 15 pregnancies, across a period of 23 years between 1993-2016, with 38 offspring surviving as of 2023. An Indian woman gave birth at 73 years old, which means she experienced about 60 years of fertility if we assume she became fertile around the age of 13. This is nearly double the life expectancy of 'pre-historic' humans and possibly four to five times the fertile years of females in that period.

Considering the lack of predators and other threats to survival, along with the introduction of mass vaccination and improved sewage systems, it is unsurprising that the human population has surged eight-fold to eight billion in just about 200 years. During the previous 300,000 years, this population barely exceeded one billion.

Humans view themselves as 'special', considering their value greater than that of all other animals. From nature's perspective and concerning the Earth's future, the opposite now appears true, as we destroy species 3.8 billion years in the making.

THE BEAT OF LIFE GOES ON

Most aspects of life can be described in terms of physics and chemistry of a level of sophistication that cannot be matched by inanimate processes. For me, this explanation is more awe-inspiring than any kind of belief that life is directed by mysterious forces beyond the reach of scientific scrutiny.

Growth in living cells played a crucial role in causing specific chemical reactions. Pasteur claimed that chemical reactions went beyond being merely fascinating aspects of cellular life; they were fundamental characteristics of life itself. Pasteur summarised this by saying 'chemical reactions are an expression of the life of the cell'.

This varying and expansive range of chemical reactions that happen within living organisms is known as metabolism. It is the basis of everything living things do, maintenance, growth, organisation and reproduction, and the source of all energy needed to fuel these processes. Metabolism is the chemistry of life.

Throughout human history most lives have ended through infectious diseases. Attacks made by bacteria, viruses, fungi, worms and a host of other parasites have claimed the lives of billions of humans, many before the end of infancy. Today, vaccines, sanitation, and medications often cure or prevent premature death.

Vaccines and medications not only prevent and cure loss of life, but can indeed lead to fatalities and health issues, a fact noted by anti-vaccine advocates, and accepted as true by the medical profession. However, the clear benefit for human survival and reproduction of vaccination is clear, as humanity has multiplied eightfold since Edward Jenner launched the smallpox vaccination in 1796.

This specific killer was eradicated from humanity, declared on November 26, 1979, about 203 years later. The spread of the smallpox virus resulted in approximately 300 to 500 million fatalities globally in the 20th century alone, and it has caused countless more deaths since it began infecting humans around 4000 to

CHAPTER 3

3000 BC. Numbers of dead for all but its last 100 years are unknown, but it is likely to have killed more humans that any other microbe including numerous outbreaks of plague. Humans may have died from reaction to the smallpox vaccine, yet this would still represent a fraction of those who have been saved by it.

*

References, sources and recommended reading or research, Chapter 3

Paul Nurse - What is Life - description of the characteristics and essentials to sustaining and developing life

Sciencedirect.com - observations on the Krebs Cycle

Smithsonian Channel - Observations of animal mimicry in survival techniques

Joseph LeDoux - The Deep History of Ourselves - observations about early Earth, evolution, human culture and the human brain

Evolution Education Outreach - observations on animal numbers and attrition rates

Berkeleyan Archive - Observations on poverty and mortality in Victorian Britain

National Post - Most human births to one female in modern era

The Guardian UK - Oldest woman to give birth to a child in the modern era

Complete Fertility Centre - observations of the effect of lack of nutrition on conception

CHAPTER 3

Ohioline.osu.edu – *Global Climate Change update*

Ncbi.nlm.gov – *genetics research*

Principles of Evolution by Natural Selection
– *BBC.co.uk*

Natural History Museum – *observations on early life*

Survivalinternational.org -observations regarding the vulnerability of isolated tribal peoples around the world

Wildlifesos.org – *observations regarding the increase in extinction endangered species caused by the expansion of humankind*

Max Planck Institute Germany – *observations of fear in babies when seeing spiders and snakes*

Junk DNA could be rewiring our brains – *University of Oxford*

Zygoma – *The unbroken chain of life*

Frontiers in ecology and evolution – *Frontiersin.org*

University of Cambridge – *observations regarding the vaccination gap*

World Health Organization – *observations regarding the effect of immunization on human population growth*

History of Smallpox – *cdc.gov.uk , Centers for Disease Control and Prevention*

CHAPTER 4

Hominins

—

As mentioned previously, our forebears diverged from the common ancestors of chimpanzees about seven million years ago. From fossils, scientists have discovered that the first evidence of Bipedalism (walking on two legs), occurred in human ancestors around five million years ago yet they were still predominantly tree climbers for over two million more years, until their grasping foot was lost within evolution.

Approximately 2.6 to three million years ago, the earliest evidence of stone tool use shows that humans were developing greater brain capacity. Approximately two million years ago, Homo erectus became the first hominin to migrate beyond Africa. Homo Erectus displayed more human-like characteristics compared to earlier hominins, including larger brains, dexterous fingers, and longer legs.

Scientists found indications that, around 800,000 to 200,000 years ago, human brain size began to grow significantly larger in that 600,000 years than over the previous two million years, probably due to improved cooking techniques that maximised nutritional intake from a wider variety of foods. The story of human brain growth closely relates to nutrition and its availability.

Humans on Earth today see our closest living relatives as the great apes, especially chimpanzees, but that was not always the case. Our ancestors shared the Earth with six different species of hominin. Some scientists believe it could have been up to eight.

The most familiar and latest of these being Neanderthals, closely enough related to Homo sapiens that the two species interbred. Modern humans still carry traces of Neanderthal DNA. Homo sapiens originated in Africa approximately 200,00 to 300,000 years ago and began migrating from the continent between 60,000 and 70,000 years ago, coexisting relatively peacefully with Neanderthals during this time.

In summary, Homo sapiens and Neanderthals had similar sized brains; however, growth within the Homo sapiens' brain appears to have had an enlarged frontal lobe within the neo-cortex, disproportionately more than that of Neanderthals.

The neo-cortex section of the brain controls reasoning abilities. Homo sapiens could outcompete Neanderthals in times of resource scarcity thanks to their superior reasoning abilities. More effective communication and

CHAPTER 4

articulating complex issues through speech and planning likely being key. Humans thrive today not due to strength or speed, but because of their intelligence.

Scientists believe that human brains began to advantageously evolve with the onset of the cognitive revolution approximately 70,000 years ago. Around 30,000 years ago, Neanderthals, who competed with early humans for food and shelter, became extinct.

HUMANS AND BACTERIA

Around 3.8 billion years after the first single cell sustainable microbes existed on Earth, evolution and natural selection has arrived at humans – if nature could feel disappointment, I'm sure it would. Humans can't survive without the ancestors of microbes which first populated Earth 3.5 billion years ago – bacteria. Surprisingly, bacteria and viruses are responsible for more deaths in humans than all other causes combined. Microbes sustain human life while also posing significant threats to it.

Efficient food digestion, sufficient to preserve human life over millennia, could not take place without microbes. Scientists believe that the number of microbial cells in the human body is comparable to the number of human cells. We primarily share a symbiotic relationship with most microbes, but we tend to be more aware of the ones that cause illness or death.

Bacterial cells, whether they dwell in a garden, the depths of a warm ocean, the Arctic, a jungle, a cafeteria at Kew Gardens, or assist in probiotic functions in your colon, must perform tasks to survive and thrive that are similar to those humans undertake: evade threats, identify and absorb nutrients and energy sources, and regulate fluids. To ensure survival of their species, they need to reproduce. Like us, they meet many of their survival needs, in part by behavioural engagement with their environment.

The ability to multiply quickly has been the foundation of success of microbes and why they have thrived in every corner of our Earth from the bottom of oceans to the tops of mountains, on and inside every living thing. Without them, life would cease to progress. They break down waste, build soils, recycle nutrients, capture nitrogen from the air that plants and animals need to grow. They were and are the foundation of multicellular life.

After one hour, starting with a single bacterium that replicates every 20 minutes under ideal conditions, there would be a total of eight bacteria. After six hours more than quarter of a million, and somewhere between hours 12-13 the number of bacteria would be a greater number than all the humans that have ever existed.

It is fortunate that optimal conditions do not support sustained linear growth over extended periods, as nutritional shortages and a toxic environment caused by microbial waste products will constrain this expansion.

CHAPTER 4

Near-perfect conditions can occur for brief stretches. The human body serves as a warm, moist container rich in nutrients and oxygenated blood. Built in systems for carrying away waste. Ready-made luxury apartments we know as cells, is about as good as it gets for microbes. Humans and other animals possess natural protection through white blood cells (leukocytes) that combat infections; however, these defences often struggle against the rapid replication of invading microbes. When they become too overwhelming they will interfere with basic survival essentials – respiration, blood quality, body temperature, metabolism – causing death.

In suitable hosts, the microbes will enjoy an abundance of resources and grow exponentially. Evolution has shaped suitability over hundreds of millions of years, which helps explain why certain diseases can or cannot spread between different animal species.

With the highly destructive strains of plague, which infected the population of Europe over a seven-year period from 1346-1353, estimates range from 30% - 70% of the human population of Europe being wiped out, approximately 50 million people. Between 1918 and 1920, an influenza outbreak in Europe, which originated from American military forces entering the continent, claimed the lives of more than 20 million people worldwide, with some estimates exceeding 30 million, primarily affecting Europe.

The outbreak was traced back to the USA, yet is referred to as the 'Spanish flu' due to its eventual identification in

Spain, which gradually diminished by 1920. Microbes, including viruses and bacteria, have remarkable survival capabilities. The Spanish flu remains in a state of suspended animation within the bodies of the deceased preserved in the permafrost of northern Europe, particularly near and inside the Arctic Circle. Some scientists anticipate that global warming will create the perfect conditions for a resurgence in the future.

The American Cancer Society has connected certain bacteria to the development of specific cancers or at least the health issues that allow the cancer to develop. Bacteria in the stomach (Helicobacter Pylori) aid ulcers which can be treated with antibiotics and wider research into this is at relatively early stages. If left untreated, though, these can lead to some types of stomach cancer.

Chlamydia Trachomatis can infect the female reproductive system and spreads through sexual intercourse. Studies have linked this infection directly to the onset of cervical cancer. This cancer rarely occurs among gay women, nuns, and other females who do not participate in heterosexual intercourse. Just two examples of how bacteria pervade our lives and play a key role in the life and death of humans.

Microbes of many kinds appear expert in survival; expert that is, in survival of the species, individual deaths in their trillions being of little importance. They have endured for 3.5 billion years and will remain the last living beings when our solar system's life reaches its conclusion.

CHAPTER 4

HUMANS AND THEIR BRAINS

'We are all atheists about most of the Gods that humanity has ever believed in. Some of us just go one God further.' Richard Dawkins, the God Delusion

'Progress is impossible without change, and those who cannot change their minds cannot change anything.' George Bernard Shaw

'Is man merely a mistake of God's? Or God merely a mistake of man?' Friedrich Nietzsche

'Every time a baby is conceived, it inherits a unique genetic composition that has never existed before. What is considered special today may well turn out to be mundane in organisms that diverge from us in future; they will have their own traits that make them different from us, and thus special in their own way.' Joseph LeDoux, The Deep History of Ourselves

In the Deep History of Ourselves, LeDoux makes the following points. If you really want to understand Human nature, we need to understand its evolutionary history. As biologist Theodosius Dobzhansky once said, 'nothing in biology makes sense except in the light of evolution'.

"We are our brains." This statement is viewed as accurate by some and ridiculous by others. What is clear is that who we are is shaped by our brains. Brains give us the ability to think, reason, to feel pain and joy, to communicate through speech, to reflect on moments of our lives or to plan and make decisions, and also worry about our conceptualised futures.

Everything we see, touch, feel, taste, smell, sense occurs in the chemistry of our brains. When those chemical stimulations, supported by electrical impulses, no longer take place, it is logical to assume, we no longer exist.

LeDoux makes further interesting and challenging points. Why do we have to look so far into the past to understand the origins of the human brain and its functions? Unicellular organisms like bacteria don't have nervous systems, much less a brain. Why can't we just focus on animals with brains, or at least a nervous system if our objective is to understand the human brain. A key argument (in the Deep History of Ourselves) is that the core survival requirements were first solved by the first successful organisms living billions of years ago, and they passed their solution to every organism that has followed.

The most primitive part of the brain is the brainstem surrounding the head of the spinal cord. This 'root' brain regulates basic functions like breathing and the metabolism of the body's other organs. This primitive brain cannot be said to think or learn; it is a number of pre-programmed regulators that keep the body functioning as it should, ensuring survival. This brain dominated in the time when reptiles were the most advanced creatures on Earth.

CHAPTER 4

Until the second half of the nineteenth century, the interrelation of living organisms was generally viewed as a progression – this placed humans above all other creatures. Aristotle's grading of nature, called the "scale of nature" or "ladder of life", was based on organisms and their degree of complexity. Humans classified as most complex were at the top. The presence of blood was considered the dividing line between vertebrates and invertebrates.

Christian theologians later developed a ranking system based on the Genesis creation story and Aristotle's hierarchy, viewing 'perfection' as one's proximity to God. Through this concept they regarded humans, created in God's image, as the most perfect organisms on Earth.

The theory was that all life began more or less simultaneously about six thousand years ago when God populated the Garden of Eden, most famously with people, apples and snakes, and created the distinct organisms that continued to exist in the same form over millennia.

A contrary view began to emerge through the writings of Alfred Russel Wallace and Charles Darwin. Darwin, though, captured the spotlight of history. Darwin argued that all living organisms are connected to one another, by way of common ancestry.

Darwin based his conclusions on evidence, scientific principles, logical arguments, and detailed reasoning –

attributes and tools that were not accessible to those who contributed to the Old Testament. His perspective was instrumental in establishing what we now accept as fact, and current opinions on how natural selection influenced evolution have changed very little. Yet some religious commentators continue to favour unproven myths in place of scientific fact.

Mid-Twentieth century writers continued to place humans at the top of 'the tree of life'. This was driven by the theory of brain evolution that viewed the human brain as the hierarchy of brains, our vertebrate ancestors, with a reptilian brain lower than mammalian brain, an early mammalian brain which in turn was lower than a higher kind of mammalian brain possessed by primates, with the human brain being at the pinnacle.

The assumption that humans are top of the evolutionary heap continues to influence views about brain evolution, as well as about the nature of the human mind and realms of ethics and morality. Humans find it challenging to give up the belief that we hold a status and superiority over all other beings.

As noted earlier, about seven million years ago, humans and chimpanzees shared the same ancestors. Subsequently, a divergence happened: the forebears of chimpanzees flourished in arboreal habitats, whereas the forebears of humans became better suited to existence on grasslands. Standing on two legs gave a better view of predators and prey, and eventually led to upright humans. Chimps did

CHAPTER 4

not need to walk on two feet in trees and jungles, leading to their evolutionary separation into different genera.

Over four million years past until what we would identify as human-like animals appeared. In his excellent book, Sapiens, Yuval Noah Harari explains, humans first evolved in East Africa about 2.5 million years ago from an earlier genus of primates, Australopithecus, which means southern ape.

About two million years ago some humans migrated to North Africa, Europe and Asia. Different environments migrated to influence natural selection, driving changes and diversity amongst our ancestors. From the jungles of Asia and Africa, to the frozen wastes of northern Europe and the deserts of North Africa and Arabia, different species evolved: Homo Erectus in Asia and Homo Neanderthal in Europe, amongst two species we are familiar with today.

Homo erectus lived for two million years, demonstrating remarkable durability and longevity that Homo sapiens are unlikely to match. From around two million years ago until about 40,000 years ago, multiple human-like species cohabited our planet.

The period from 40,000 years ago being fairly unique rather than the norm we consider it today where Homo sapiens exist and the next most intelligent animal being one of the great Apes. The Earth once supported up to six species of human-like creatures at the same time.

Some commentators have placed this figure as high as eight different species of humans.

The earliest humans of 2.5 million years ago had a brain size of three times the average mammal of the same weight, whereas modern human brains are six to seven times the size of an average mammal of the same weight. The neo-cortex in humans is six times larger than that of a chimpanzee, our closest living ancestor. Large brains carry a significant cost; human brains account for roughly 3% of body weight while consuming about 25% of the body's energy. They consume three times more than the brains of our closest relatives.

For over two million years, the human brain continuously developed whilst making only nominal progress, such as development of basic weapons and tools. For much of the last two million years humans were down the food chain, hunting small game and being hunted themselves. A step change in the development of humans occurred 800,000 years ago with early and occasional use of fire. Around 300,000 years ago, Homo erectus and neanderthals used fire daily for warmth, protection, and cooking activities.

Cooking opened access to foods that were previously difficult to digest such as wheat, rice and potatoes, foods we consider staples today. It also sterilised foods such as meats which could carry microbes and parasites dangerous to humans. Humans expend approximately 10% of their caloric intake on digestion, but this percentage may have likely been much higher, when they processed raw food before discovering fire.

The difference of intelligence between the evolutionary line of chimps and humans is being exacerbated by chimps expending many calories on hours of chewing food and many more digesting it. Humans, through cooking food, spent far fewer calories on digestion, enabling more calories to be used on the brain's development through 'leisure time', defined as functions unrelated to survival. The future basis of arts, music, science etc.

Cooking makes digesting fruits, nuts, insects, and meat much easier. The chimpanzee, our closest relative, spends about four to six hours each day chewing its food to obtain the necessary nutrients for survival and thriving, while a human typically spends less than one hour doing so and still maintain an excess of nutrients, reflected in excess bodyweight in many modern humans.

Fire served a dual purpose; it not only aided in cooking but also created a greater divide between humans and other animals. Animals rely on their size, strength, speed, and stealth to gain an advantage, while fire allowed humans to harness a significant power. Humans might clear forests and light fires at night to safeguard themselves from predators while maximising the nutrients obtained from their food. A game changer, fire was a limitless resource.

Humans began hunting larger game between 300,000 and 400,000 years ago, reaching the top of the food chain around 70,000 to 100,000 years ago, although some estimates put this date much earlier. However, at a stage in human history our ancestors progressed from hunted to hunter.

Approximately 70,000 years ago, a pivotal shift occurred in human history: the Cognitive Revolution began. Scientists remain uncertain about all the specific factors that accelerated the development and functionality of the Homo sapiens brain, giving it an advantage in reasoning over Neanderthals. Homo sapiens developed the ability to think in innovative ways and to express themselves with complex language, enhancing both their communication range and precision.

Further development of language led to humans being able to communicate about things that do not exist as well as things that do. God, Limited Companies, the themes of the Magna Carta and American Declaration of Independence, Laws etc. These ideas exist solely in human minds, but they influence our behaviour and shape our imagination. We record them with fluid on paper, in language, now on computer, all generated by the human imagination. Language could be used to describe, with accuracy, and one of the main influences on modern Homo sapiens is the supernatural, especially religion.

In a fascinating insight into the differences between humans and other animals, LeDoux explains, these imagined and human conceived ideas (myths) have encouraged cooperation in the human species – religious

CHAPTER 4

people who have never met, can cooperate against a common theme, sometimes in opposition to different religions.

The same principle holds true for followers of political factions, football fans, and individuals from the same nation; borders exist only in the human mind. A myth forms, propaganda develops around it, leading to its acceptance as truth and the growth of its followers.

The key lies not in sharing the story to convince others, but in getting them to pass it on with the same faith and enthusiasm. It would be difficult to create states, churches, companies, political parties if we could only use what we know exists. Modern society depends on myths for its existence.

The belief in Gods is a human concept which given the apparent arrogance that surrounds it, 'we are right and non-believers are wrong' can be a considered demeaning and insulting to our ancestors who didn't have the brain capacity to conceptualise a similar type of God.

Humanity approximately between 10,000 to 6000 years ago had advanced far enough in the Cognitive Revolution, to not only conceive that beings which have no physical form can exist, but were able to convince themselves and others of this existence. Later, not only that they existed, but that they were far more important than those beings known to exist. Had no Gods been conceptualised until recently, the reaction to any contemporary human introducing the theory that an invisible being was more important that our children, would be laughable.

Religion before family became a common theme. Rose Kennedy, the mother of John F Kennedy, when hearing her daughter Kathleen was to marry a divorcé with children who was not Catholic, disowned her daughter. When Kathleen was killed in a plane crash shortly after, Rose commented 'that God had taken divine retribution'. It is a sentiment that appals me.

German philosopher Ludwig Andreas Feuerbach believed that humans created God rather than the reverse. If evolution rather than creationism is factual, then it is clear that at a time in the past when evolving from animals and earlier versions of humans, Homo sapiens started the belief in God.

There is credible evidence to suggest signs of religion related faith were displayed between 45,000 to 200,000 years ago – burial with personal artefacts being a sign of basic religious practice. On balance of probability this is more likely to have been more prevalent following the Cognitive Revolution of 70,000 years ago. However, there appears little sign of the exploitative nature of religion which became prevalent with large civilisations and the creation of countries, and the use of religion to control behaviour of the masses.

Belief in God(s) began to form in the evolving human mind, and whether it was 6,000, 10,000, 70,000, or even

CHAPTER 4

200,000 years ago is not fully relevant to the fact that there was a start of God, within a human brain. That brain which previously could not conceive of a God. I have opted to focus on developed religion with millions of followers, specific Gods rather than small numbers of hunter gatherers possibly practising religion through grief, superstition or similar.

Some will therefore argue that humans had the capacity to conceptualise Gods before 10,000 years ago. Without knowing for sure I concede that it is possible; however, we can generally agree that at a time between the start of the cognitive revolution and 6000 years ago, humankind developed a brain that could conceptualise invisible, possibly non-existent beings that were beyond the capacity of their ancient ancestors, in a similar way that a motor car, mobile phone or space telescope were beyond the capacity of the ancestors of modern humans.

The human brain's capacity to conceptualise a God would not have developed and taken hold in a short period of time. It is likely that human hunter gatherers, within their groups, may have envisioned a supreme force or spirits, probably not a God as modern humans describe them.

Not all human brains function equally; some people process concepts more effectively than others, as observed in contemporary humans. Gods and non-living spirits would have been conceptualised maybe thousands of years before the first recorded gods. Yet still relatively recently, when compared to the total time presence of humankind on Earth.

In smaller groups, people might have shared any new thoughts with only a few individuals, while later societies had much larger communities that facilitated the spread of ideas and new practices. Farming communities had several thousand years to potentially develop their ideas about gods before larger communities later emerged.

Larger communities relied on co-operation, trust, individual and specialised functionality, within community living. As people enjoyed more leisure time, storytelling gained popularity as a favoured hobby. Stories that stretched the boundaries of human imagination, infused with mystery and hard to decipher, yet plausible for a curious mind, would flourish and gain popularity. Wondrous acts of gods or mythical beings would have been popular.

I found no reference to an omnipotent, invisible, all-powerful God, similar to the Christian God, to my knowledge, in the first 290,000 years of Homo sapiens populating Earth, the Homo sapiens, without whom, humans would not exist today. In the 60,000 years after the onset of the human cognitive revolution, enhancements in brain chemistry, nutrition, physical size of cognitive sections, and neural connections led to a gradual increase in human brain capacity, enabling people to conceive of God in a more intricate manner, especially over the past 10,000 years.

CHAPTER 4

In summary, farming enabled larger communities, created more leisure time, and raised nutrition levels in times of plenty. Human brain growth had been accelerating at a faster rate for 60,000 years, since the start of the cognitive revolution, from 10,000 to 6,000 years ago the human brain had a potency it had never had before. The invisible, possibly non-existent could now take control of the human brain, through storytelling and propagation of ideas. Becoming as an instrument of control by the human brain, of other human brains was an inevitability.

A striking example of using religion and God to control the masses was the 'Act for the Advancement of True Religion' in 1543 which prevented the masses from reading the Bible. Access for the masses was limited to listening to religious leaders reading passages in church, with the interpretations, conclusions and controls that were so easy to apply to followers. Most followers had little to no formal education, which made them susceptible and easily influenced.

ABC Science, Jennifer Vegas and Discovery News highlight that Monks from 15th century England would hoard food, much to the disgust of the English middle classes. Peasants and villagers often provided food to the monks, seeking God's favour, believing that by nurturing his perceived representatives on Earth, they could ensure their own good fortune in the afterlife. This whilst commonly suffering from malnutrition, poverty and plague.

Archaeologists discovered signs of subsistence living and malnutrition in the skeletons of food providers, while they also found instances of obesity and type 2 diabetes in the consumers. In addition, other conditions related to obesity were also discovered, hyperostosis, arthritis and weight-related back problems. Clear evidence of mind manipulation of the poor, impressionable and uneducated.

Humans advocate for, and spread ideas, that they believe in, including concepts such as God, religion, nationalism, and politics. Some will even kill others or take their own lives, not because they are definitively and provably right, but because they believe they are. That belief, contained in just a few pounds of flesh within a hard casing, that is so easily manipulated by exposure of its senses. The only link the human brain has within the world beyond is via its five senses; if these can be corrupted, then the individual brain can be corrupted.

God is consistently used to explain the, as yet, unexplainable; in addition 'he' is credited with favourable outcomes as well as disasters and tragedies, which are for a reason in 'his name'. 'He' can do no wrong.

Religion has prevented honest and detailed 'self-analysis' by humans over millennia. How can a perfect species made in the image of God, be accountable for the action of God? The Will of God is responsible for everything, even though 'he' then judges the use of free will. This allows God to be credited and those who oppose 'him' held accountable. It seems a jumble of self-serving nonsense.

CHAPTER 4

Religion has undoubtedly prevented or slowed scientific discovery, over centuries, as many scientists discovered, some paying with their lives. Making superstition and ancient writings dominant over evidence demonstrated the promotion of ignorance. The cosmos, evolution, our ancestry, Earth's recovery after disasters and the insignificance of humankind within the Universe, hidden by the church because these discoveries did not suit its misleading narrative.

Let us pay tribute to Giordano Bruno, philosopher, poet and astrologist, in 1600AD burned alive for heresy by the Spanish inquisition. His crime, to claim that stars were like our Sun with worlds like our own planets, surrounding them. Religion, the rule of incredible ignorance.

Sight and sound serve as the brain's most powerful senses for influencing decisions and therefore our initial relationship with the concept of God. Humans draw upon information from their senses to influence their emotional decisions throughout their lives. For example, psychological torture often targets these senses in addition to inflicting physical pain.

Deprivation of sight and sound has emerged as a highly effective method of torture, utilised by various government security forces and terrorists globally, to obtain information. The deprivation of senses can have a devastating impact on human emotions, potentially breaking the human spirit. Individuals have been broken to the extent of, not only giving up information but in changing religion and beliefs, such is the impact.

Few humans claim to taste or smell God(s), many say they can feel God(s), sight and sound having absorbed information from others that this is a possible human feeling. As God(s) are invisible with no physical presence in some religions, the use of the touch sense will not be used to feel God, it is therefore an emotion. 'I feel I will be with God, one day', 'I feel God is with me' are very different contexts to 'I am not going to touch the hotplate, as last time I burned myself'. Two statements driven by emotion, the other by reflective learning from physical experience. 'I feel' being a stimulus of the brain rather than a statement of physical contact.

God, good, evil, heaven, hell, limited companies, laws, money and millions of other human propositions, can be conceptualised and therefore exist, but only with a level of reasoning, that on Earth, only the human brain possess. This additional brain capacity has also been used to develop schools, battlefield tactics, healthcare, fairgrounds, GP surgeries and everything else humans have created in the modern world. No animals would give up a carrier bag full of food for a few pieces of paper.

NUTRITION DRIVEN FAITH

For seven million years, ever since our split from the great apes, humans lived as hunter-gatherers. The agricultural revolution began approximately 10,000 years ago. While this presented challenges for humans, it triggered a

CHAPTER 4

substantial population surge. From roaming hundreds of square miles, learning where food would be and when, humans started to grow food and domesticating animals. Significant time needed to be spent on clearing land, using fire on many occasions, and protecting food from animals and other humans.

As societies relied more on farming, they also became dependent on weather conditions and crop yields, which led to hardships from famine and increased conflict over land ownership that became essential for survival. Humans often had to make a choice, between some of their number starving, to preserve seeds for the following year, or all starving as no seeds were left for planting the following year.

In today's world, people continue to engage in armed conflicts over land, particularly areas that are fertile or rich in resources such as oil, metals, minerals, and water. Land provides food, which offers resources for reproduction and population growth, leading to greater strength in numbers, and of course, wealth.

In Mein Kampf, Adolf Hitler considered eastward expansion as a chance to grow the Aryan population while also aiming to eliminate communism and others considered undesirables. Land can serve as a buffer zone between opposing forces, providing an early warning and allowing time to prepare for possible aggression. Expanded population and resources are viewed as the way to secure further and stronger power. Land continues to play a crucial role in ensuring human security and survival.

Farming, whilst eventually enabling massive expansion of the human race, caused a great deal of suffering when it went wrong. This is now forgotten as the currency of evolution is not hunger, desperation or pain, but the propulsion of DNA into the next generation.

The agricultural revolution allowed a greater number of humans to survive compared to what hunter-gathering could have sustained, although often under significantly harsher conditions. Without it, the Earth's human population would probably be many hundreds of times smaller than today's eight billion inhabitants.

Food availability has a strong connection to human reproductive capabilities, as it does for all mammals. In times and regions where food is abundant, young females tend to reach puberty sooner to facilitate higher reproduction rates, given that sufficient food supports a growing population.

This is achieved by varying hormonal levels which delay the ability to conceive longer in times of food shortage. In concentration camps during the Second World War the ability to conceive amongst those on a low-calorie existence was greatly reduced, preventing pregnancy from rapes often imposed by captors or fellow inmates. To become pregnant could have meant death of both mother and unborn child.

Farming led to a significant rise in the human population, growing from approximately eight million during the

CHAPTER 4

hunter-gatherer era to around 250 million by the time of Jesus Christ, and residing in increasingly larger communities, including towns and cities. Population size was controlled and regulated through disease within close-proximity communities, where it spread more rapidly, and wars between communities.

The advent of efficient sewerage and vaccination in the 19th century and beyond within concentrated populations, for example in cities, released nature's limiters on human population growth, to lead humanity to the eight billion people it has today, with many living subsistence, undernourished lives.

The growth of farming was seated in the ambition of not only surviving but achieving economic security. The growth of human communities ended this ambition – those working the land were soon subject to the rise of the elite. Chiefs, royalty, religious leaders, emperors and politicians ensuring food surpluses and more were removed in taxation, or better described as the use of strength to extort the weak. Government officials, priests, artists, thinkers, and the military spent their time doing other things but still needed to be fed and sustained. The price of organised civilisations and new cultures. Farmers were at least afforded more protection, as those extorting also relied on the food supply.

With this civilisation came the myths as discussed earlier. In Christianity, a widely held belief is that 'we are all equal in the eyes of God', yet some people often enjoy more privilege than others. The truth is that evolution is not about equality, it is about achieving difference; equality is a human myth, not practised at all by animals and rarely by humans. Evolution and the difference created, is the essence of survival. Socialist and communist politicians are often 'feathering their own nests' instead of using resources to equalise society.

Whilst humans continue to strive to have more than the next human, equality remains a myth. Six generations or more, of socialism, have not repressed the instincts built over 20,000 generations in humans or socialists, despite its admirable intent. Daniel Coleman raises related points in his book 'Emotional Intelligence' which shed light on the deeper aspects of the human brain.

Despite social constraints, passions overwhelm reason time and again, as the emotional brain dominates the practical brain. The essence of human nature stems from the fundamental structure of mental life. Human emotional design we are born with is what worked best for the previous 50,000 human generations, not just the previous 500 and certainly not the previous five.

On a related point, Michael Shermer points out that humans are a hierarchical social primate species, who, despite centuries of democratic rule, still long to sort themselves into pecking orders within families, schools, peer groups, social clubs, corporations and societies. We can't help it; it is in our nature, courtesy of natural selection operating in the social sphere.

CHAPTER 4

The slow deliberate forces of evolution that shaped our emotions have done their work over the course of millions of years; the previous 10,000 years, despite having witnessed the rapid rise of human civilisation and explosion of human population from as low as five million to eight billion, have left little imprint on our biological templates for emotional life.

Better jobs, housing, vehicles, food, money are all strong drivers in contemporary humankind. They derive from thousands of generations when survival relied on having more than the next human, more food, shelter, protection, land or weapons. Instinct is not changed over a handful of generations, only the currency has changed. Having more than the next human was, in the past, an opportunity to raise the chances of, and sustain, survival.

Every individual possesses a unique genetic code and experiences diverse environments, upbringings, values, and belief systems. These factors set humans apart and create inequality among them. Society relies on myths like the belief that everyone is equal. These myths unite society until people stop believing them. The myths of Roman, Greek, Norse and many other gods ceased to be believed in human minds, as they were replaced; therefore these gods (myths) no longer exist. Human myths are disposable, unlike DNA, individual instincts and respiration.

The myth of God being an eternal entity, existing outside the human brain, is certainly not supported by history. God appears to have existed since human brains could conceptualise gods and when they did the were very different from the gods of today. Numerous imagined religions have come and gone, and all will eventually come to an end. Contemporary humans refuse to believe this, yet Norse, Romans, ancient Greeks and Egyptians, were once contemporary humans.

When humans die their brains die with them, life will continue to exist, especially microbes, gods will die when the last human brain that believes in them dies. God seems to have emerged in the human mind when humans were finally able to imagine a God, but not earlier. People have shaped their understanding of God in a remarkably influential and widespread manner.

Whilst belief in God holds significant importance for many people, it does not play a role in evolution. Myths do not connect with survival in the way that DNA is passed down. Infants cry when they feel hungry, adolescents experience urges to reproduce during puberty, and everyone needs to find shelter for survival; all these instincts are encoded in DNA. Understanding laws, how limited companies work, timetables and conceptualising God are not.

The need to belong appears to be instinctive in humans, which leads to conformity with the views of others. And it continues from generation to generation. It is the need to belong, security and complying which has found its way into human DNA coding, not the need for God and religion. Varieties of humans have had these needs for millions of years, it is only lately they have manifested themselves into modern, human-conceived God and religion.

CHAPTER 4

Unique human cognition and reasoning, developed by building on the cognitive capabilities of our ancestors. We need to understand our cognitive abilities to understand their origins.

Cognition is used for thinking, reasoning, planning, deciding and similar. Cognition has played a crucial role in shaping the understanding of human nature before civilisation. Cognition enables inner awareness and self-reflective consciousness, inner awareness of self as essential stages of the thinking process.

The adding of religion enabled large numbers of strangers to cooperate as they believed in a common myth. Any large-scale cooperation, whether modern state or ancient city, is rooted in common myths that exist only in people's collective imagination. Churches are rooted in common myths. Two Christians who have never met, can go together on a crusade or pool funds to build a new care home or church because they both believe in the same god.

THE INFLUENCED AND MANIPULATED HUMAN BRAIN

On 1 August 1966, Charles Whitman took an elevator to the observation deck of the University of Texas building, in Austin. The then-25-year-old started firing indiscriminately at people below. Fourteen people were killed and 31 wounded, until Whitman was finally shot dead by police. When police got to his house, they discovered he had also killed his wife and mother the night before. It was a surprising, random act of violence, seemingly out of character, a lack of anything about Charles Whitman that would seem to have predicted it. He had, however, left a note; extracts include:

> *I don't really understand what compels me to type this letter. Perhaps it is to leave some vague reason for the actions I have recently performed. I am supposed to be an average reasonable and intelligent young man. However, lately... I have been a victim of many unusual and irrational thoughts... After my death I wish that an autopsy would be performed on me to see if any visible physical disorder...*

Whitman's request was granted. The pathologist reported a small tumour had been found, the size of a small coin, and it was pressing against a part of the brain called the amygdala, which is involved in fear and aggression. This small amount of pressure on the amygdala led to terrible consequences in Whitman's brain, resulting in him taking actions that would otherwise be completely out of character.

His brain had been changing and 'who he was' and what he believed, also changed. This is an extreme example, but less dramatic changes in your brain can alter the fabric of who you are – the consumption of drugs or alcohol, for example. Types of epilepsy make people more religious. Parkinson's disease can often influence people to become compulsive gamblers. It's not just illness

CHAPTER 4

or chemicals that change us; from watching movies to the jobs we do, listening to those who influence us, everything contributes to a continual updating of the neural networks, we summarise as 'us'.

David Eagleman in his book 'The Brain – the story of you' further substantiates these points. He says, our thoughts and our dreams, our memories and experiences all arise from this strange neural material. Who we are is found within its intricate firing patterns of electromechanical pulses. When that activity stops, so do you; when that activity changes due to injury or drugs you change in lockstep. Unlike any other part of your body if you damage a small part of your brain, who you are is likely to change radically.

Eagleman reasons that, in the coming years, we will discover more about the human brain than we can describe with our current theories and frameworks. At this moment we are surrounded by mysteries; many that we recognise and many we haven't yet registered. Only one thing is certain: our species is just at the beginning of something, and we don't fully know what it is. We're at an unprecedented moment in history, one in which brain science and technology are co-evolving. What happens at this intersection is poised to change who we are.

For thousands of generations, humans have lived the same sort of life cycle over and over; we're born, we control a fragile body, we enjoy a small strip of sensory reality, then we die.

LeDoux adds, our everyday language arose and is sustained because it enables discourse about inner lives of people as they interact with others. It has been extremely advantageous for humans to have this ability. Without it, human culture would not exist.

Science may give us the tools to overcome traditional evolutionary barriers. We can influence our own hardware and as a result our brains don't need to remain as we inherited them. We're capable of inhabiting new sensory realities. Eventually we may even be able to lose our physical form. Advances made in just a few decades has set us apart from our ancestors in the degree of control we have over our own lives and its outcomes. Humans are now discovering the tools to shape their own destiny. Who we become, is up to us.

When we know more about the human brain it will be possible to engineer changes, through drugs, surgery, or impulses, possibly convert a devout Christian to an atheist, an atheist to a devout Christian, a pacifist to a serial killer, changing personalities of any named type to another. Can you picture future despots similar to Hitler, Stalin, Mao and Pol Pot becoming volunteer charity workers? In future, this will not be as far-fetched as it now seems.

Joe Kennedy, father of President John F Kennedy, had his daughter Rosemary lobotomised in 1941, as he believed she was promiscuous and an embarrassment, also a threat to the reputation and ambitions of the Kennedy family.

CHAPTER 4

A small part of Rosemary's brain, part of the frontal lobe, was removed in an attempt to remedy this issue. As a result, she was institutionalised from 1941 until death in 2005, living with a mental age of a five year old. A vibrant, funny, intelligent young woman had been reduced to a life-long dependent child, by a few incisions into her brain. All humans would be vulnerable to this level of intervention.

A crude operation by today's standards, the introduction of microsurgery delivered by skilled surgeons, development of drugs, mind altering torture or propaganda have advanced the human influence on the human brain. These techniques are developing rapidly; the ability to manipulate and influence the human mind, will increase significantly in future.

Humans have been manipulating the brains of other humans for thousands of years, without drug-induced or physical alterations. The use of propaganda and, at times, torture have been used, by those with an agenda of control. Born in Brazil, on balance Catholic, born in the Middle East, Muslim, born in India, Hindu or Sikh, all examples of cradle to grave mind manipulation. Willing participants, who would, I'm sure, object to my summary of their lives, yet the evidence is compelling. Religious faith instilled as unquestionable fact from birth, is likely to have that effect.

Totalitarianism has generally failed in the modern Western world after short periods of thriving. War or internal revolution typically leads to its defeat, but the future possibility of manipulating the human brain physically, along with past psychological methods, offers possible opportunity of changing brains and eliminating internal resistance. Where democracies would be unlikely to introduce compulsory physical brain alteration, extremist governments are more likely to view this as legitimate and effective policy, when the technology is available. Yet another instrument of control of the human mind by those who wish to control.

Our emotions are central to what we are and become. The workings in the brain appear fascinating and complex, not least because of their role in mental suffering. Our ability to understand how any psychological process works in the brain is only as good as our understanding of the process itself. We need to know what we are looking for in order to find it.

As stated earlier, humans have survived and expanded, not by being bigger, faster, or stronger, but by being cleverer, using reasoning levels which have never existed before. Unlike many organisms, we did not merely evolve by adapting our bodies to the changing world; instead, we utilise our 'human' abilities to transform the environment around us. We can 'load the dice of life' in our favour and make more environments advantageous to our bodies and way of life.

No other animal, not even our nearest primate relatives, possesses the capability to conceive ideas such as

CHAPTER 4

constructing a palace, discovering a cure for an ailment, composing a sonata, or writing a scientific thesis; then explain the concept to another person, strategise its implementation, and subsequently implement it. Human cognition is unique, but this does not mean we are better or more entitled than our ancient ancestors' offspring, with which we currently share the planet.

PHYSIOLOGICAL CHANGE

Our bodies and our brains change so much during our life, the changes can be difficult to detect. Every four to six months or so, human blood cells are entirely replaced and skin replaced every few weeks. In approximately six to eight years, every atom in a human body will be substituted with different atoms.

Humans undergo ongoing physiological renewal from conception to death. The process of renewing loses efficiency over time and becomes less effective with imperfect replacements; this phenomenon is known as aging. Fortunately, there may be one constant that links all these different versions of yourself together – memory – but even this will degrade with age, and likely cease on death.

Our species has experienced a remarkable transformation over the last 100,000 years; we have evolved from primitive hunter-gatherers living on meagre resources to becoming a hyper-connected civilisation that shapes its own future. Today we enjoy mundane experiences that our ancestors could never have dreamed of.

There is a significant drawback: our bodies are made of physical material. They will age, deteriorate and die. At the end all your neural activity will cease and then the experience of being conscious will come to an end. It is the fate of us all. In fact, it is the fate of life, but only humans are so foresighted that they concern themselves. We can conceptualise time both before life and after death; this is likely to be unique to humans.

THE HUMAN BRAIN, A THING OF WONDER, A THING OF NAIVETY

We are our brains. As humans we have several parts of our brains that are similar to our mammalian relatives and others that set us apart. All our bodily functions are in place to support the function of the brain, and with it, realisation.

For example, our brainstem regulates breathing, heart rate, blood flow, motor and sensory pathways (nerves), alertness and sleep, all basic functions of mammals. Our cerebellum regulates balance, learning, emotion, co-ordinate movement and attention. Our thalamus, spatial attention, depth perception, acts as a relay centre for information, from the body to the brain, consciousness

CHAPTER 4

and alertness. All actions coordinated in an instant, by unconscious intelligence. DNA having pre-programmed much of this.

Then there is the neo-cortex, the 'game changer'. All we see today that is human-made, houses, cars, pens, computers, phone, clothes, everything came from this 'box of tricks'. It has also organised conflicts (wars), medicine, religion, good and evil, excessive greed, a need to dominate through psychological means, planning and empathy amongst many other traits unique to humankind on our planet.

Human cognitive thought processes and mental models, conceptualising, reasoning and language are products of the kind of our human brain, within the brain is to a large degree a product of the cerebral cortex, especially in the neo-cortex.

The neo-cortex has four main parts – occipital lobes responsible for sight and processing visual information; parietal lobes responsible for depth perception, spatial orientation, receives sensory input, language processing, spatial attention, writing / reading and calculation; temporal lobes memory, combining senses with memory, object recognition, understanding language, art/ music, speech and hearing; and then the jewel in the human brain crown, the frontal lobes, these are responsible for personality, social skills, judgement, emotional regulation, movement, speech, REASONING and executive function (cognitive control).

Many of the functions within the neo-cortex are common in other mammals, for example wolves, apes, lions and many others use 'speech' when communicating, a basic form compared with humans. Our ability to reason allows us to learn and cultivate significantly more intricate forms of communication.

All of these animals can make loud sounds that may be heard within a few miles, humans can speak to the other side of the planet or even into space within milliseconds through technology developed by human reasoning. The ability to reason, plan our actions, calculate the consequence of our actions and deliver an outcome using various resources truly sets us apart from our closest ancestors, the great apes.

LeDoux states that 'goal directed instrumental learning' is sometimes talked about in terms of natural selection, at the level of the individual. In other words, trial and error learning enhances fitness by selecting adaptive behaviours in a single animal, much like natural selection enhances the fitness of species through genetic selection of adaptive body traits but on a multi-generational level.

Behaviours can be learned when they have beneficial outcomes, such as obtaining sustenance when energy or fluid levels are low, or preventing harm. Behaviours can be varied and unpredictable within humans, lacking a specific connection to a goal, which is crucial for demonstrating behavioural flexibility. The explanation for why behaviours that achieve such goals are learned is that they have emotional outcomes and impacts.

CHAPTER 4

The ability to link complex thought with learning, was a significant change in the behaviours within the animal kingdom. It has enabled the capacity to base responses on memories about consequences of trial and error learning.

Deliberation is the consideration of response choices, use of knowledge stored in memory to make the decision that seems most likely to produce a useful outcome. Additionally, through deliberation, various connected and unconnected pieces of information can be obtained across several steps of reasoning to reach a favoured action, sometimes in milliseconds.

What would usually require many repetitions of trial-and-error behaviour for a new response to be learned via outcome-based reinforcement, can be converted into internal simulation of the outcomes choices.

Aldous Huxley noted 'that it is in language we have raised ourselves above Brutes'. The human capacity for language is essential to, and a unique component of, human cognitive mental life. However, deaf people and people who lose speech because of brain damage are able to communicate and participate successfully within society.

It's not the ability to talk that is key, what matters is what underlies talking, what language does for cognition. In 'Kind of Minds', Daniel Dennett, philosopher, stated: "The kind of mind you get when you add language to it is so different from the kind of mind you can have, without language, that calling them both minds, is a mistake.

Language lays down the tracks on which thoughts can travel".

The Greeks attempted to classify nature and the natural world. Language was the key. Language allows thoughts to progress in different directions, and yet stay connected. It provides words which enable the labelling of external objects and to characterise and recognise our thoughts, perceptions, memories, concepts, emotions, beliefs and desires.

The words humans use reflect the things of significance in their culture. For example, Benjamin Whorf made famous the notion that people living in snowy environments have names for, and can recognise, more kinds of snow than those not living under such conditions, because snow is important for their ability to survive and thrive.

Language serves more than just the purpose of naming and categorising objects and events; it also structures their underlying thoughts. With language also comes grammar, which structures human mental processes and informs when we are thinking, planning, and deciding.

The cognitive scientist Edmund Rolls commented that grammar allows humans to plan actions and evaluate their consequences by anticipating many steps ahead, without actually having to perform the actions. Most other animals, Rolls notes, are limited to innate programmes, habits, and rules. Or, in the case of mammals and birds, reinforcement-based learning.

CHAPTER 4

Non-human primates have enhanced cognitive capabilities and skills in reasoning and solving problems. But without the ability to use complex language into deliberation, thought remains static and low level when compared with humans.

Animals can communicate with one another about specific things. Birds can attract mates, recognise their young in large social groups, mammals summon partners to outnumber threats or competitors. Complex language becomes the ability to use sounds or visual symbols flexibly, to represent ideas about the present, past, or future; only humans have that capacity.

Only humans can use language to communicate to other humans exactly when and where a particular kind of danger exists and then discuss implications with others for planning ahead to alleviate that danger. Through language, humans boost capacity to engage in conceptual and schematic thinking, beyond mere observation.

British anthropologist Robin Dunbar proposed that early social benefits of language involved not just the sharing of knowledge about sources of nutrition and proximity of enemies, but also about individuals, enabling gossip. Our early human ancestors used language to communicate details about who they could trust and the qualities that good partners possess. He observes that social interactions often involve gossip among humans.

Evan MacLean, University of Arizona, argues that what separates human cognition from other primates is our ability to resist competitive impulses and engage in cooperation, reason about intentions and desires of others of our kind, and to communicate with one another using complex language. This capacity is not guided purely by genes but depends on accretion of incremental learning across generations, human culture.

Rakoczy and Tomasello observed that a human child living alone from birth on an island, lacking the experience of learning that includes cultural history, may have the cognitive ability more like a chimpanzee than an adult human. Genes are insufficient for shaping the human mind; we also require history and culture.

This is very different from a spider or a crocodile where genes alone are sufficient for life's necessary and initial functionality for survival. This does get supplemented by learning as life progresses, but not to the extent of humankind.

Celia Hayes concluded that culture has played a crucial role in separating humans from the great apes and making possible what humans have achieved throughout our history. The significant role genes play in the natural selection in the process, remains a key contribution.

LeDoux states that human reasoning, mental models, pattern processing, conceptualisation, language and the like are products of our human brain. Within the brain, cognition is to a large degree a product of our cerebral cortex, with a major contribution from the laterally-located neo-cortex.

CHAPTER 4

Because human working memory capacities are greater than in other primates, scientists believe the large size of the prefrontal cortex in humans explains the cognitive differences and abilities. When measuring brain size differences between animals, the most widely used calculation in recent times has not been absolute volume or weight as brain size increases as body size increases. Instead, most researchers use brain size relative to body size.

An elephant's brain is three to four times larger than a human brain. An average adult elephant typically weighs between 4,000 and 6,000kg, making it 60 to 90 times heavier than the average human. The additional brain size clearly doesn't support a greater cognitive capacity than humans possess, but is used to support the additional functionality to support an animal of such size.

This is viewed as an estimate of how brain function remains after basic function parts are accounted for. Findings indicate that humans have a comparatively larger prefrontal cortex than other primates, correlating with the cognitive benefits unique to humans. Human cognition can be susceptible to malfunction in conditions, for example autism and schizophrenia.

Humans, with no alternative experience are likely to prefer the life we lead, but in the end, there is no effective measure, other than the ability to survive, that can measure whether a human life is a better or worse kind of life, than that of apes, horses, cats, dogs, fungi, plants, archaea or bacteria. Our human brains tell us being human is preferable; any other animal is unlikely to be able to consider alternatives.

If species' longevity is the measure of success, we will never be better than ancient single cellular organisms. Therefore, if the importance of a species on Earth is measured by longevity, the ability to survive, then it is bacteria and not humans who are the chosen species. Humans would not exist without them. Bacteria would continue to thrive without humans.

Predicting and modelling our future is a particularly important skill, as it makes it possible to go beyond options supplied to us humans by natural selection, and/or instilled in an individual by goal-directed learning.

Humans' biological entities, chemicals put together in structured way, whose constituent parts operate using complex cooperation, similar to all life. Some humans have developed a 'rogue component', a brain network that can choose to undermine survival and purpose of the rest of the body. This appears unique to humans. This is the phenomenon that underlies human consciousness, and especially our capacity for autonoetic (the ability to mentally model oneself in relation to time), reflective self-awareness.

For more than two billion years, unicellular organisms dominated the Earth. Multicellular organisms evolved fitness and survival ability from a single cell to a more complex entity with many cells that all shared a common genome. The biological model worked well for another

CHAPTER 4

billion years or more, until harmony was suddenly challenged by the arrival of the capacity for autonoetic-conscious brains in humans.

Humans were unremarkable compared to coexisting primates, then at some point between 200,000 and 50,000 years ago (probably around 70,000 years ago), something happened to further distinguish humans from other primates. We developed new capacities and ways of existing and communicating with one another, language, reasoning, self-representation.

The unique structure of human brains allows people to defy natural selection and the way life on Earth has operated for 3.8 billion years. We can change our environment to suit our needs better and while also shielding ourselves from fears and dangers. Imagining the unknown and 'yet to be' inspires humans to find new ways of fulfilling their lives.

Our thirst for knowledge has led to scientific and technological discoveries that have made life, at least for the fortunate among us, easier in many ways. Most humans do not face the peril of hunting for food in risky environments, as ferocious predators typically do not exist in their everyday lives. We easily adapt to seasonal changes in temperatures with clothing, weather proofed shelter and convenient appliances. We have access to medications to treat, and even prevent, illnesses; and surgical procedures that can fix, and in some cases, replace damaged body parts.

With our autonoetic-conscious minds, we have constructed conceptual guidelines, such as morality, religion and ethics, to help make decisions about our way of life.

Mammal brains have many parts, each responsible for different functions. Many of these are necessary to sustain life such as heartbeat reflex. Many functions we need but can't control consciously. These have been built by DNA passed by our ancestors, a chain of codes forming messages of 'construction' similar to the instructions that built our stomachs, hearts, feet and every physical and mental attributes we are born with.

The vast majority of the human brain is very similar to mammalian relations – we all need heartbeats, oxygen in our lungs, feelings of hunger; in short, the ability to survive and thrive. The neo-cortex being the separator, being responsible for everything we see around us from farmed fields to skyscrapers, brutal wars to hospitals, pens to computers. Also, many of the things we can't see such as Gods, Demons, Father Christmas and Ghosts.

THE HUMAN BRAIN AND GOD(S)

Humankind's cognitive abilities have created a thirst for knowledge, discovery, exploration and asks questions such as 'why are we here', 'what is our purpose', 'is there a supreme guiding power'? For millennia the greatest

CHAPTER 4

influence on the human mind and behaviours has been God(s) and religion. This relationship deserves close examination; any intelligent extraterrestrial observer witnessing the evolution of Homo sapiens from the beginning would likely regard the emergence of belief in the supernatural as a significant shift in the status quo.

The human neo-cortex can envision a God and transmit the story of God across generations. Our relatively recent ancestors, around 10,000+ years ago, did not appear to have, at least not in the way modern humankind has developed belief.

The evidence suggests that humans did not have the brain capacity to conceptualise invisible, omnipotent gods until relatively recent history of Homo sapiens. We can be confident that ancient man had no symbolism to the modern gods; however, drawings of animals in caves suggests a possible spirituality, in the respect paid to food and predators.

Prior to 200 years ago, humans would have been unlikely to conceptualise television, modern communications, skyscrapers, aeroplanes, spacecraft except for the most forward-thinking humans such as Leonardo da Vinci, but even this was in trying to emulate bird flight rather than a passenger jet carrying hundreds of people.

Five thousand years ago, humans had not invented the wheel, and it was unlikely to have been conceived at that time, more than 1500 years before it was invented, initially for making pottery. That pottery wheel changed the world when adapted to vehicles and today the use and reliance on the wheel is worldwide.

Gods have been all consuming for many humans, stimulated by those cognitive powers locked inside the brain. Sacrifice of loved ones, even suicide in the name of a God have demonstrated how the human brain can be convinced and manipulated by external influences and internal emotions. These cognitive abilities also motivate many individuals to question the notion of divinities created by humanity.

The human brain traditionally struggles to accept new information that goes against established beliefs. While beliefs do not equate to knowledge, the human brain tends to prioritise them over knowledge that contradicts those beliefs.

A child coming to terms with the 'loss of Santa Claus' clearly demonstrates this. At first, the child wants to deny the truth as the lie is comforting, familiar and secure. I remember being told by children at school that Santa didn't exist. I returned home, hoping my mother, the person I trusted most, would tell me the other children were wrong, but the new information was confirmed as knowledge. I therefore in that instant realised that previous information, giving by those I trusted most, was false. Told for good and compassionate reasons, but false, similar to how I later viewed human conceptualised God(s).

CHAPTER 4

Humans are informed from an early age that myths are true, by those they trust most, 'therefore myths must be truth'. If my mother had told me that the children at school were wrong, I would have believed her, to later have my faith in her rocked, when the truth emerged.

The opening up of the information highway, ability to share information and indeed scepticism and science has led to more and more questioning and doubting. When transferred from religion to politics, the reason many totalitarian governments restrict use of the world wide web.

The breaking down of barriers that once stopped us investigating the clergy and making it answerable to law, previously answerable only to God's judgement in the eyes of many. At last believing testimony of those offended against by those who used God and religion to satisfy their personal greed and lust. This leading many to believe that a key purpose of religion, as practised by many, is control, power and wealth for a few. Its benevolence often being a smokescreen to hide its real purpose, to dominate and control.

For example, the Catholic church sought to control sovereign states across Europe for hundreds of years through the use of God. In addition, in modern times, it has used its power to cover up the serious wrongdoing of the priesthood, placing the reputation of the church ahead of the needs of the vulnerable. Many sovereign states showed deference to the power of the 'Holy See', thus enabling religion to control state and Law and perpetrate wrongdoing, without challenge.

The conflict between control and benevolence within religion has been an ongoing issue over centuries. Comforting bereaved individuals, then sentencing those with differing views to excommunication and even death. Making homosexuality a crime against God, and with it, sentencing between 10-20% of followers to misery, guilt and unfulfilled lives. Tens of millions sentenced to a life in fear, believing they were sinners, having to be covert and secretive regarding their sexuality, in many cases prevented from participation by religious affiliation.

Faith is a good thing and should be separated from religion. What right do a number of differently dressed men, women or others have to claim faith, belief in something beyond this world, for themselves or the organisation they represent? To tell others how to lead their lives, in the name of a mere concept of the human brain.

I am fortunate to be a middle-class man and living in a first world country. The majority of my fellow human beings have much less materially than I do, and the vast majority of those will always have much less. Many people in poverty possess strong faith. Where would they be in a life without faith, in a world where they continually struggle, are surrounded by poverty and grief. They have hope that conformity within this life will make life in the world beyond life will be improved – how likely would conformity be, if that hope did not exist?

CHAPTER 4

Myself and my fellow middle-class sceptics such as Richard Dawkins etc sleep comfortably in our beds because of religious faith, amongst the impoverished masses. Without it, there could well be tyranny by those who have little, against those who have much. It would be difficult for many to accept that their under-privileged hardship of existence is as good as life will ever be. Not to risk the wrath of God in the hereafter, may well free up human minds to live their lives differently. Without belief in God, the tipping point of anarchy amongst the oppressed, may have changed significantly.

Richard Dawkins refers to the 'selfish human gene' as a good thing and as far as survival and humankind arriving into the 2020s, it certainly has been. However, without the control of 'good' and 'evil', creations of humankind, religion, also a creation of humankind, and thereby the self-proclaimed, exclusive purveyor of 'faith', the human 'selfish gene' may have led to revolution and the destruction of world order as we know it, many more times than has been attempted to date.

The same selfish gene played its part in creating those human behaviour controllers to give advantage to the ruling classes – the concepts of good, evil and religion focused the human mind and its behaviour.

While the 'selfish gene' has played a crucial role in the survival of humankind by creating advantage and therefore improving survival changes, it may gradually contribute to our extinction. Can we foresee a reversal of greenhouse gases within the economies of China, USA, Indonesia, India, Brazil, the five main polluters, Africa and so many others dependent on creating greenhouse gases as part of an economic growth process to release countries from relative poverty? Unlikely, or at least unlikely within the time it must happen to make a real difference. Many countries have started the process but slowly, watching what others do before committing sufficiently.

Who cares about future generations when there is money to be made and power acquired, goodwill targets are set then changed, hedonists rule the world. The 'we'll be dead when it happens' brigade are in charge. China's ambition to surpass the USA as the largest economy currently appears to overshadow the future of billions of other global residents, including future Chinese people, who might endure the repercussions. When one economy disregards the rules, others will likely feel encouraged to do the same and maintain their economic competitiveness.

BATTLES FOR THE HUMAN MIND

A great many human beliefs through millennia have been highly influenced if not fully determined by won and lost battles, or fear of being ostracised or worse by those who have gained the power of life and death over other humans.

For example, if Roman Emperor Constantine had not defeated his co-Emperor Maxentius in 312 AD, having

CHAPTER 4

converted to Christianity, would the Christian faith have survived? The mostly likely answer, is yes, but as a minority religion of which there were several at the time; thereafter, it would likely have perished, as many religions did.

In 324 AD, Constantine became Emperor of a fully reunited Roman Empire after he defeated his co-Emperor Licinius at the battle of Chrysopolis. Christian followers grew from six million to 33 million between 300-350 AD, mainly as a result of two battles. Both victorious battles were credited to the Christian God.

Many ancient battles often featured inflated accounts of victory told by the winning side, including exaggerations about enemy numbers and casualties. This decoration amplified the stories of successful generals and attracted greater admiration from their followers. Perhaps embellishment, or simply lying for gain.

In 333 BC, during the Battle of Issus, Alexander the Great triumphed over the Persians; some eyewitness accounts reported that the Persians outnumbered the Macedonian and Greek forces by a ratio of 8 to 1. Contemporary studies analysing the evidence estimate this at a ratio of 1.6:1. Embellishment was common, even expected; a number of Greek and Roman victorious battles had similar embellishment added by the victors.

Many different achievements were exaggerated by the 'winners' throughout history. It is estimated by modern historians that Christianity had fewer than 10,000 followers at the end of the first century, more than 60 years after the death of Jesus Christ. At this stage Christianity could have been classed, not unreasonably, as a minority 'losing faith'. Yet within 200 years its following had grown to six million, and just 50 years later it had a following of 33 million, more than 50% of the population of the Roman Empire.

This suggests that the push for Christianity became more prominent in people's minds between 100 AD and 300 AD. In 312 AD, the Battle of Milvian Bridge marked a decisive turning point for Constantine and Christianity. In 324 AD Constantine as sole Emperor of the Roman Empire promoted the word of the Christian God. Christianity was now victorious, a 'winning' religion, embellishment of stories to support the spread of Christianity part of normal practice and encouraged by the leader of the western world, plus some of the eastern world.

Christianity has depicted Christ as a white European. There can't be any doubt that he was of Middle Eastern appearance with Hebrew / Arab features of dark skin and a pronounced nose with dark eyes, as his followers would have been. If 12 apostles did exist, they would have had Hebrew names, yet these would have been changed by the Christian religion to suit the European image placed on them. Abraham, Caleb, Aaron, Isaac etc would have been likely male names of the time, certainly not Matthew, Mark, Luke, Peter, John etc.

Pre-Islamic religion and gods was made up of a mix of Iranian polytheism, Judaism and Christianity, the latter

CHAPTER 4

being a product of Judaism. The influence of the two monotheism religions presumably influencing the single god, of modern-day Islam.

An accurately depicted Christ would very likely have resembled a Moorish Saracen, more than a white European and likely to have been a Galilean Semite according to Richard Neave. Little doubt remains that Jesus Christ would have had dark skin, eyes and hair.

This image would have been unacceptable for 11th to 13th century Europeans, who fought the religious wars of the crusades. It would have been unimaginable for Christian Kings and Popes to have fought for an individual who resembled and spoke similarly to their sworn enemy, yet that is undeniably what happened. The naming or renaming of the apostles with non-Hebrew names, appears a further attempt at anglicisation or Europeanisation of Christ.

The image of Christ changed for the convenience of those in power. Christ being a deity from the fourth century would appear to have suited those in power, wanting to impose their will and beliefs on the masses.

The son of God, walking on water, feeding 5000 people using a few pieces of food, bringing the dead back to life, curing leprosy without antibiotics and coming back to life two days after dying. Not actions that would attract fewer than 10,000 followers in 60 years, but actions that would attract 33 million people, 300 years after the death of Christ, later to be billions, with considerable embellishment and promotion from State(s) and church.

Stories and legend worthwhile forsaking their own gods for, and pleasing their Emperor, who had God-like status. By 350 AD Constantine the Great had been dead for 12 years, two of his Christian sons had succeeded him consecutively and Christianity was established as the main religion of the Empire, with the potential to spread throughout the Western world. A very different story from that of 33 AD, 100 AD or even 300 AD.

The Roman religions which Christianity replaced, are now obsolete and their gods likely to be used as quiz questions, but never again for worship by the masses. Much of this stems from a couple of decisive battles that shaped historical events. The future of Christianity had rested on battlefield tactics and the changed beliefs of one man.

Currently, 2.4 billion Christians live on the planet, representing 30% of the global population. Had Christianity been defeated, holy men of today would be espousing the merits of the 'winner' and its updated variations. Christianity, very likely, wouldn't exist and humans of today would be oblivious, as they are today about Canaanite, Atenism, or Manochaeism religions.

People in Europe, the Americas and beyond still live with consequences, good and not so good, of the Battle of Milvian Bridge. When we discuss battles that changed our world, Issus, Hastings, US War of

CHAPTER 4

Independence, Waterloo, Kursk, Midway etc, these pale into insignificance, in lasting impact, when compared to the battles of Milvian Bridge and Chrysopolis.

Consequently, this provided a substantial advantage for the forthcoming supremacy of Christianity across the Western world. Christianity in one form or another has spread to the four corners of the globe. This would have been improbable without the support of the most influential person on Earth from centuries ago.

Many people have died in the name of Christianity whether in religious wars or persecution and witch hunts. Humans' brains being humans' brains, if it hadn't been Christianity another religion's set of beliefs would have become dominant and many would have also died in its name. Clearly demonstrating the futility of human justification for murder in the name of religion.

The irony of the reformation in England will never be lost on my human brain. The split of the Catholic Church and the seeds of Protestantism demonstrates, to me, how the human emotional brain so easily fools itself. One of the most significant differences between the two sub-religions stems from whether a person believed in salvation being realised through faith and good works, the Catholic view, or salvation being achieved through faith in Jesus his atoning sacrifice, the Protestant view.

Both of these beliefs rely on the existence of good and evil; to atone for sins and seek salvation you need to cleanse your 'soul', or so religious leaders will tell you. Good and evil are myths invented in the human mind, similar to laws and scout membership. None exist in the animal kingdom as they are not necessary for survival.

Henry VIII serves as an example of a king who manipulated religion to fulfil his own ambitions. Unable to divorce his first wife, a devout Spanish Catholic, because the Pope would not dissolve his marriage, Henry abolished the Catholic Church as the official religion of his kingdom and embraced another faith. The impact on people's lives has been highly significant since then, thousands have lost their lives and millions have endured suffering for the desires of one man. Individual humans exploit God and religion to serve their own interests.

Any change of religion within a society is likely to take at least one, but more likely a few generations. The human psyche is so strong and the willingness to admit previous error, or change views, stubbornly resisted. Change of any kind can remove human emotion from its comfort zone. Beliefs instilled from birth become hard to challenge once they are accepted as the unquestionable truth. Parental, state and church influence have been at work.

Transitioning from the Roman, Greek, and Norse gods to Christianity took time and likely led to objections and divisions, as some people adopted new ideals while others held on to their familiar beliefs. Ultimately, the combination of authority, numerical superiority,

CHAPTER 4

persuasive propaganda, and the fear of being different would have secured victory for the newest religion. People adapt their beliefs quickly when they see obvious benefits, often replacing traditional Gods with more convenient and less dangerous alternatives.

In contemporary and future times, followers of religion will inevitably migrate to science but this won't be as definitive; it is possible to be both religious and believe in the wonders of science. However, the inconsistencies which appear to exist between the two will inevitably place science as an influential vanguard for the future of humankind.

Francis Bacon, arguably the father of science, died about 400 years ago; the influence of science was at first diminished against that of religion, but thanks to other pioneers and those prepared to stand up for what they knew to be the truth, it now thrives. Few should doubt evidenced scientific findings which appear to contradict religious beliefs with conviction, beyond superstition or dogma. But many do, their brains having been saturated in irreversible propaganda from birth.

Imagine science had made the discoveries about the Universe that it has, over the past 30 years, before religious commentary then announced that the Universe was made in six days, a snake, an apple and a few ribs involved in human creation. How silly would that appear? Yet we still allow the writers and promoters of this fantasy to influence our children's lives. Why in being misled ourselves, would we want to mislead our children?

In the past 200 years, science has changed the lives of humans significantly through sanitation, travel, medicine, communications and other game changers. Religion, although sometimes well meaning, caused the loss of tens of millions of human lives, in its name. Science has taken away lives through weaponry, but its net contribution to humanity has been positive, especially through vaccination and sanitation. If I was in the vicinity of a smallpox outbreak in 1800 and had the choice between Edward Jenner's vaccine or going to church to pray, it would be an easy decision.

The influence of science over the past 30 years has been to change many views on creation, the Universe, to humankind's relatively small, almost insignificant part in time and space. All factual contradictions of the Bible.

In August 2023 the Catholic church announced acceptance of openly gay people, having previously and for centuries proclaimed it as a sin against God, though many in the church continue to believe it to be. I wonder to what extent loosening the grip the church has on people's minds, has led to this decision. It appears to be targeted at maintaining popularity of a diminishing church than maintaining traditional and entrenched values. A tipping point was reached and the church had to take painful and divisive action. Now going a small way to promoting what a growing majority of people have known to be morally correct, for decades.

The church has the same amount of right to comment on two humans copulating or conducting their lives as

CHAPTER 4

it has on the habits of sheep or spiders. None. It is more evidence of the ridiculous need for power and control by a few over many.

It appears to have been of little importance to the Catholic church, that tens of millions of gay Catholics lived unfulfilled lives, riddled with guilt over many centuries, yet when its popularity and powerbase was diminishing, change could be made through a papal decree, disguised as moving with the times. The church now abandoning its centuries-old foundations in exchange for relevance.

Increasing the capacity of the human brain, enhancing access to credible information, transforming that information into knowledge, and fostering open discussions across nations will likely diminish religious adherence. The evidence of shift is clear in many Western countries, possessing these freedoms. Separation of church and state has been instrumental in enabling this significant shift.

The difference can also be viewed when comparing Christian countries where religious teachings are strongly promoted by the State. In traditional Christian countries such as Brazil, Colombia and Peru, where education levels are relatively low 70% of people believe in God. Compared with the UK, Spain, Canada and Australia where under 50% of people believe in God, and in Sweden under 25%. Significant differences exist in wealth, access to education without heavy religious influence, and length of access to the internet, are all a factor. In 2022 84% of Brazilians had gained access to the internet, a decade earlier it was half that number.

Brazil has the largest number of Catholics of any country in the world, but this has seen a drop of 29.3% since the 1970s from 92% to 65%, Protestantism has absorbed some of those, accounting for a growing group who believe in God and Christianity, but reject Catholicism. However, many more now reject religion. As access to information, scientific discoveries and alternative, potentially evidenced views, there is no apparent reason why religion shouldn't continue to decline.

Some forecasts are for growth in religious affiliation over the coming decades. I remain to be convinced; however, faster growing populations in less developed countries, with poorer education standards and little access to alternative views, will assist this higher forecast. It is likely that nations with limited educational opportunities and reduced access to information and alternative views may see an increase in religious belief.

Developed countries with higher standards of education combined with more freedom of expression are unlikely to experience growth; the reverse is likely, tempered by global migration from underdeveloped countries.

In conclusion, typically higher birth rates in less educated, impoverished countries may account for a positive growth forecast. The impoverished and less well educated appears to supply a solid foundation for increased religious following.

CHAPTER 4

As covered earlier, good and evil appear to be inventions of the human brain. Before humankind took up this moral crusade, to further regulate or control other humans' behaviour, good and evil didn't exist. We don't consider a hippopotamus killing another hippo in a fight for parenthood to be either good or evil, yet a human killing another human is usually viewed as evil, unless, of course, if it was done in the name of religion or its lovechild, politics. Good and Evil, right and wong being in the 'eyes of the beholder'.

Homo sapiens have inhabited Earth for approximately 300,000 years. Religion, along with its ethical guidelines and its mind regulating frameworks, likely dates back less than 6000 years. Religious concepts of good and evil likely developed in the human brain over roughly two percent of Homo sapiens' existence.

Put another way, less than 0.00016 of the time, life on Earth has existed. Prior to that these 'good and evil' actions were just part of existence, human killing human as accepted, and with as much accountability as a hawk killing a smaller bird or a lion killing cubs that don't carry his genes, to bring female lions into oestrous, and enable him to father cubs earlier. No animal in the history of life on Earth has drawn up codes to regulate behaviour, beyond instinct, survival of individuals and of species were the only codes.

Good and evil has led humans to unfairly label snakes evil serpents, spiders and even goats as representatives of evil. One human writes a book using myth or superstition, Dracula for example, and suddenly bats become an increased target of suspicion – only in the human brain could this happen. These animals, similar to humans, have been adapted over millions of years to compete to survive, they stand at the furthest extreme of the evolutionary branch within their species, it is the height of ignorance to label them otherwise. Christian and other religion has played its part in this injustice, whilst many other religions respect animals and their spirituality.

For millennia, the human brain has yearned for knowledge, particularly regarding the rationale behind human existence. It seeks a clear understanding of life, but only theories have surfaced instead of definitive answers. Theories abound and the longest surviving is based around God and the interpretation of God (religion).

Religion is possibly the 'answer' humankind has sought. It is, however, a convenient answer as it explains the otherwise unexplainable. We then have the dichotomy of this highly complex and thinking part of the human brain that can only go so far in explaining 'what it's all about'. The rest is added by the 'emotional' part of the human mind; these explanations are generally along the lines of religious ideas which manifest themselves differently around the planet, different perceived myths.

With different religions all claiming they are correct, as their members have been indoctrinated to that conclusion,

CHAPTER 4

religion is confusing, especially the 'thought fascism' which seems to accompany it. The Romans, Greeks, Norse and many other followers of now extinct gods, once thought their gods were the only gods and teachings passed down as undeniable truth. Religions of their time, as major religions are of today, but these modern religions do not represent religion of the distant past and are highly unlikely to represent religion of humans, in the distant future.

The Muslim religion, mainly split between Sunni and Shia, is a powerful example of one major subcategory of a mainstream religion believing something that is contradictory. Sunnis believe the Prophet Muhammad declared a successor and Shias believe this to be inaccurate and believe in a different line of succession.

Given these two beliefs are incompatible, one sub-religion is living under an illusion, but both will say, not us. This different view has caused animosity, suffering and death for nearly 1400 years and continues to divide followers of Islam today. Has anything been gained by this difference of opinion? It would be difficult to make a case for a positive outcome of this difference.

Reasonable (true) and unreasonable (false) is in the perception of the beholder and within human groups, will have existed in human groups well before organised religion on a mass scale, but probably driven by non-religious forces. Self-preservation and protection of the group is a very strong driver in mammals; animals protect their pack and offspring through instinct, sometimes at considerable risk to themselves. Religious followers appear to have inherited these instincts.

Unity consistently proves to be more potent than division among mammals, within groups, enhancing survival prospects for both the group and its individual members. Nature has ensured survival of the species rather than survival of the individual, made one dependent on the other, and this is hardwired into animals, especially mammals.

I am certainly not making a case that morality and the human belief in good and evil is all destructive; it is one of the main drivers of civilisation and maintaining order as we know it. We do, however, need to understand, that good, evil and morality itself, is a very modern human concept. Used for the positive in forming reasonable, compliant behaviour within society, but nevertheless, the concepts have also been misused for the purposes of controlling humans and steering their actions for personal gain of another human(s). Introduction of sin, fear, manipulation, guilt, for self-interest.

Russia's conflict with Ukraine serves as an example, fuelled by the belief that it represents a battle against evil rather than just an imperialistic desire for land and resources. In the 1940s, the Nazis made similar claims, positioning themselves as defenders against what they viewed as the Jewish and Communist 'menace'.

Humans uniquely feel disgust, emotion, or outrage when they witness atrocities on television happening to

CHAPTER 4

strangers in other countries. Yet similar reaction and action assaults on family or pack members is something that will usually be met with violence or defence by many mammal and bird species. It is in the 'here and now' and of concern; a similar attack in a bordering territory or on another continent was, and is, of no consequence.

Morality is a force some humans turn on and off as it suits personal aspirations and objectives. The concept of morality being of primacy in wars is an irony, though some attempt has been made through conventions or rules of war. The German army which invaded Russia and the Russian army which defended, 1941 to 1945, demonstrated little sign of morality, or distinguishing between good and evil. A war of annihilation, where no rules existed, precipitating torture, murder, rape and other extremes of human depravity.

Similarly, the East Anglian witch hunts of 1644-1647, in the name of God and religion, where women who were deemed to have signs of witchcraft were executed usually after being tortured into confession. It was a way of subjugating and controlling women. Owning a cat, having scars or spots, a facial wart, having a birthmark or possessing knowledge about natural cures for illnesses could condemn an innocent victim to torture, hanging and occasionally being burned alive. Victims were starved, beaten and tortured into confession. Where religion, or its name, sank to one of its lowest points and so clearly demonstrated ignorance, discrimination, indoctrination and wilful murder. Where religion met Nazism, it has on many other occasions.

I perceive no difference between the workings of the indoctrinated human minds that participated in mass exterminations in Nazi death camps and those which participated in the extermination of innocent women in East Anglia. Whether political or religious doctrine, the effects on the gullible human mind are similar. In this comparison it is only volume of atrocity which is different.

Chapter 5, verse 32 of the Quran, Surah Al-Ma'idah, states:

> *'If any person slew a person… it would be as if he slew the whole people'.*

For the individual who dies, the whole world does end.

Stimulus within the minds of those who carried out these atrocities being similar to Nazis in Russia, and the Spanish Inquisition across southern Europe. All believed they had right and/or God(s), on their side, this justified the actions in the minds of the perpetrators.

Yet it was only a stimulus in their human brains brought about by propaganda and indoctrination, similar to all religion, extreme politics and cult following. Morality is tailored by humans to suit their objectives and need to gain advantage, a clear sign that the 'well meaning' parts of religion, politics, morality don't change two million years of evolved instinct. The same areas of the human brain are stimulated when a dictator, religious

CHAPTER 4

leader, cult leader or similar is taking control of human minds through espousing ideas, established through propaganda, usually infused with lies and distortions.

These instances of 'good' and 'evil' brought about by propaganda, the master of morality, whether religious or political. The human brain processed, indoctrinated to the extent that it totally believed it was on the side of right or good rather than wrong or evil. These are powerful examples of the ease with which the human mind can be manipulated, especially if you indoctrinate from birth or from a position of individuals disadvantage by promising advantage.

Religious indoctrination, cult influence, and political conditioning of the human mind seem almost identical. It is truly unbelievable what a human will do to another human because a third party has informed them that it is right and justifiable. The human brain is the most advanced brain on Earth, yet it is incredibly easy to manipulate. Advanced the human brain certainly is, but that extra cognitive power comes with it, a need to belong, a need to believe, a need to be correct and a resistance to changing views once the 'seeds of perceived knowledge have grown inside it'.

*

References, sources and recommended reading or research, Chapter 4

The Deep History of Ourselves - Joseph LeDoux - observations on human culture and the evolution of the human brain

Sapiens - Yuval Noah Harari, observations on human evolution and the development of human culture

The God Delusion - Richard Dawkins

Emotional Intelligence - Daniel Goleman

The Brain - David Eagleman - observations of the workings of the human brain

The growth of Christianity - Rodney Stark, Humanscience.org

Quote on taking human life - The.Ismaili

Constantine's conversion to Christianity, overview and impact - Jennifer Lorenzetti & Jessica Whittemore, Study.com

Bmcgenomics.com - human divergence from chimpanzees

Smithsonian Institute, humanorigins.si.edu - Human use of tools

CHAPTER 4

Australian Museum – Homo sapiens cohabiting Earth with other humans

University of Southampton – Homo sapiens cohabiting Earth with other humans

Bacterial Community Variation in Human Body Habits Across space and Time – E k Costello, C L Lauber, M Hamady, N Fierer, J I Gordon, R Knight, National Library of Medicine, National Center for Biotechnology Information.

Newsinhealth.nih.gov – Humans need bacteria

Cuemath – bacterial division

BBC.co.uk – bacterial division

The Black Death 1346 to 1353 – Boydell and Brewer -National Archive, archive.gov – Spanish Flu 1918 to 1920

Science Daily News – cooking with fire

Princeton University – cognitive revolution

Techtarget.com – mob mentality

Evolving Brains, Emerging Gods, Early Humans and the Origin of Religion – Ludwig Feverbach, plato. stanford.edu

How God Changes Your Brain - Andrew Newberg MD & Mark Robert Waldman

How Evolution of the Human Brain led us to God - Zocato Public Square, E Fuller Torrey

Fat Jolly Monks had Painful Secrets - ABC Science, Jennifer Vegas, Discovery News

Are There Beliefs to Die For - The Guardian UK

Tragedy of Religion Stifling Science - Cape Cod Times, Stephen Pastore

Religion has Slowed Down Human Advancement - kialo.com

Torture using Sensory Deprivation - Dignity, Danish Institute Against Torture, fact sheet 19: sensory deprivation

The link between nutrition and fertility in women - Oxford Academic, the ESHRE Capri workshop group

Agriculture and increased human population - pnas.org George R Milner and Joseph L Boldsen

Neo-cortex - sciencedirect.com

Human Reasoning: The Psychology of Deduction - Jonathan Evans, Stephen Newstead and Ruth Byrne

CHAPTER 4

The Brains Autopilot Mechanisms Steers Consciousness – Steve Ayan

Functional and clinical neuroanatomy of morality – Manuela Fumagalli and Alberto Priori, Oxford Academic

Pope declares zero tolerance for Catholic church abuse – cnn.com>Europe

The Roman Catholic church investigation report – iicsa.org.uk

Why did Henry VIII break from Rome – Royal Museums Greenwich

Decline of Catholicism in Brazil – Catholic Herald

The Sunni-Shia Divide, Council for Foreign Relations, cfr.org

Depravity was contagious – New York Times, 1936-45 Nemesis, Ian Kershaw

Religion as an influencing factor of right wing, left wing and Islamic extremism – National Library of Medicine / National Centre for Biotechnology Information

Speciation and bursts of evolution – Chris Venditta & Mark Pagel

The Battle of Milvian Bridge, summary, outcome & significance - Britannica

Autonoetic Consciousness, an overview - Science Direct

The Gaia Principle - Harvard University

The seven stages of Human Evolution - BYJU's, India

Understanding complexity in the human brain - Danielle S Bassett & Michael S Gazzaniga

The vulnerability of the human brain - Christine Bulteau, Science Direct

How conformity can be good and bad for society - Zaid Jilani, Greater Good Magazine, Berkley Education

***Secular, Religious dynamics and their effect on humanitarian norms compliance**, Olivia Wilkinson and Emma Tomalin, Journal of Human Rights Practice, Oxford Academic*

Cohesive Societies: Faith and Belief, examining the role of faith and belief in cohesive societies - Madeleine Pennington, British Academy, Cohesive Societies. The Faith and Belief Forum

Human Genome editing - World Health Organization

CHAPTER 5

Realisation and Consciousness

—

We humans have a realisation, we have had it since some time before birth when our foetus made the transition from oblivion to consciousness. We can't imagine not having a consciousness or realisation because to achieve that we would need to imagine 'nothingness'. Not blackness, nothingness. Yet if we cast our minds back to before we were conceived, we can get a better idea.

The human brain is a human's realisation and consciousness. The circulation of blood and nutrients, along with the detoxification functions of other organs, ensures that life continues and prevents it from transitioning into death; 25% of our energy usage is dedicated to this process.

The relationship between the human brain and the human mind is a subject which leads the vulnerability of the human brain, even prominent scientists, to believe they are different. How can the wonderment we experience, deliberation, consciousness and spirituality, be singularly produced by a few pounds of flesh? My simple answer, is, because it is. There is no logic that can possibly rationalise that any form of consciousness or existence, can exist in humans without oxygen and nutrition through blood supply.

We can explain how DNA and chromosomes work, and their dependence on a chemical relationship and support from bodily functions such as respiration, we know that death prevents the continuation of this process. Why is it so difficult to accept that the mind is destroyed when the brain is destroyed? Do we believe that every living thing with a brain has a separate mind and soul that continue after death? They are made of very similar chemicals and as such will be in the same state in death, as Homo sapiens. We are not special, we just declare ourselves as such.

The development of the human neo-cortex to its advanced state led humans to mistrust this conclusion or rather an inability to comprehend non-realisation, in death. We can consider a possible future after death only because we can conceptualise it. Our advanced brains have created a post-death, favoured reality. Heaven, hell, purgatory and other myths.

Given the brain capacity of non-human animals, it is unlikely that they fully comprehend death for themselves, in the way humans do. They observe death of others and

CHAPTER 5

are alert to dangers through instinct, and if they live long enough, through experience. They may find it difficult to imagine death, which in concept form is unimportant to survival, but they instinctively know to preserve life as long as possible.

Non-realisation appears to be what is before and after, either side of life, and life the small interlude we are granted by evolution and ancestors' DNA, in between those great timeless voids. Why do humans have difficulty accepting the concept of non-existence after death, likely, on balance of probability, to be a very similar experience, or rather non-experience, to that before life?

Hundreds of billions of human realisations have existed, are existing or will exist in future, and in the animal kingdom that number will be almost limitless. Add to this the likelihood that trillions of planets in our Universe have life and complex life and the number is approaching infinite.

Using the smallest estimate of stars in the Universe, only one star in 20 billion needs to support one planet that supports complex life to ensure over a trillion planets with complex life in the Universe. See the (Frank) Drake equation later in this book which contests that one in 20 million stars may support complex life, 1000 times more than my example of one per 20 billion.

For the Milky Way to contribute proportionately to that trillion-complex life-producing stars, and planets, that would represent about five to 20+ planets similar to Earth in our galaxy. Inhabitants will live, they will have realisation, they will die, then they don't then have realisation. Life and realisation isn't the special or exclusive entity that many humans believe. The Drake equation estimates a much higher 10,000 planets in the Milky Way which sustain complex life at the level of humans and beyond.

Humans and their advanced brains seem to need reassurance that death is not the end but that suggests that life was not the start. Faith in the latest human-described God, even by billions of humans, no matter how passionately they believe, does not change the logical and likely outcome for us all.

Let's assume for a moment a realisation after death exists, the likelihood is that this would be unrelated to our current earthly life; our memories are stored in brain tissue which is now dead. If I have lived before in this Universe or in another dimension, I have no memory of that realisation, no loyalty or love for anyone left behind. I continue to love the memory of my dead parents in this life, but I don't expect to continue that love beyond this life. I would need a cognitive realisation related to my life on planet Earth to enable that. Once my brain is dead this is not a high probability as far as the balance of probability goes.

In 13.8 billion years I have had one realisation that I know of, it's all I know or will ever know. If realisations are reusable, I believe them to be as a blank personality canvas

CHAPTER 5

populated by new DNA when conceived. If a human was born a year after I die with the exact same DNA as mine, a clone maybe, it would have different character-shaping experiences and therefore a different person with possibly the same instincts, but it wouldn't be me.

If realisations exist beyond our lives, without the brain tissues that holds our memories, learning etc then they are a surely a blank canvas, ready to be repopulated by learning etc. An unborn baby develops to the point of realisation and beyond, there is an instant when 'it is', transitioned from not existing to existence. We know realisations will exist after we die, animals with their associated realisations have and are likely to exist for possibly hundreds of millions of years, single cell creatures probably billions of years, each individual clinging to just a few hours, months, years or decades of life.

A realisation comes from life and partners it. Parts fall off us all our lives, skin and internal cells being renewed by DNA, using nutrition as its fuel supply. By the time a human being dies, if we live an average length of life, many times our cells and tissues that supported our early and mid-life have already died, decomposed, made food for bacteria etc. Yet we mourn the relatively few cells that die at the end of a life, or rather we mourn the passing of a realisation 'with its memories'.

A shovel full of coal is likely to have the same level of realisation as the ashes of a cremated human, both nearly pure carbon.

Realisation is highly likely a blank canvas at the time of conception. Base instincts seem to be added before birth, through DNA, which carry into life. Communication skills other than instinctive such as crying when hungry in young humans, learning etc are a blank sheet. Everything we absorb into our brains, is surely lost on the death of the human or animal, and its brain.

For those who believe humans have memories after death, consider the following. Memories are held within brain tissue which deteriorates over time and loses memories in life. Humans only hold a fraction of their lives within their memories and as they age these are lost as the efficiency of their bodies and minds declines. People with dementia fail to remember people close to them and happenings that were once at the forefront of their minds – how could these memories possibly be carried into death?

Imagine the Earth had limitless resources and no disease or wars had ever existed. Every time a childbearing woman was fertile, she had become pregnant and given birth following each ovulation. With limitless resources and sanitation, numbers of people would be measured not in billions or even trillions, but septillions and upwards. Each would have a realisation and a life of learning, experience and have a story to tell. We don't mourn for them. A couple with three children don't mourn the fourth child they never conceived, yet its potential for life existed with its own story and hopes never to be realised.

CHAPTER 5

Realisations, although theoretically limitless, are, however, dependent on available resources and sanitation. Without sufficient food and sanitation, disease, suffering early death were a feature of civilisations. Many billions of humans are yet to exist, they will exist, we don't know yet who they will be, in the same way we are alive today, yet 100 years ago we were unknown. Future parents may not themselves exist yet, the opportunities for combined genes to form new humans, in future, almost infinite. Each different combination, a different person, so many billions of trillion possibilities will never exist.

Over 100 billion humans have existed, with their realisations and their story to tell, possibly hundreds of billions will exist in the future. On a planet with limited resources, lives, and their stories, need to be lost for others to take their turn. All life and realisation is temporary and easily replaceable.

Humans who lived before them significantly enhanced their chances to experience life in the future. Mass food production and transportation, sanitation, vaccination, all conceptualised and delivered by human brains, all real and physical, unlike many invisible, non-existent concepts which now run alongside.

When we die, is it important whether something of us exists? For nature, the important issue is that blank canvases can be populated and that DNA of a defined species exists to populate them and create new life, not mourn old, or extend realisation post-death, for no apparent purpose of nature. This is the way of evolution, which has been incredibly successful for 3.8 billion years.

Having advanced reasoning abilities that allow us to conceive of afterlife myths does not grant us precedence in death over all other living beings that exist, or have previously existed. Why would I be different in death from a single bacteria which lived for 20 minutes then perished 1.8 billion years ago? Indeed, 1.8 billion years from now we will both be in the same non-existent state, unrecorded and not mourned, in that present.

When an animal becomes extinct, its DNA blank canvas is lost, nature does not mourn, another species which could not adapt quickly enough has gone, natural selection now concentrates on adapting existing species to new environments to ensure survival of the fittest. A DNA set lost provides an opportunity for another set of DNA to be expanded. Should humankind destroy itself, it will be no more than a hiccup to evolution. It has been through more significant losses.

Why should death, non-existence, oblivion be a bad outcome, because the human brain fears it? How can oblivion possibly be something to fear if we won't exist to feel fear? Didn't we exist in nothingness before our conception? Advanced, but questioning, needy, insecure, pliable, easily influenced and programmed to favour the outcome it wants, the human brain is a strange entity. For all its genius in creating the wondrous world around us, the human brain is impressionable, superstitious and also vulnerable.

CHAPTER 5

To the inanimate world there would be a very long eternity in human time terms; the irony is that time, as we who live know it, doesn't exist to inanimate objects, the same way time won't exist to us when we are dead, and we have no realisation. We will become atoms and elements once again, our true permanent state. Our minds, returned to oblivion.

'What's it all about, Alfie, is it just for the moment, we live?' – Burt Bacharach and Hal David

Time appears to be an element exclusive to the conscious living who can conceptualise it, and measure its passing. The exception to this assumption could be DNA and natural selection – DNA has a dimension of intelligence as it changes its chemistry over time to respond to environmental changes.

DNA demonstrates a form of intelligence without consciousness, as humans know it, thereby realising time unconsciously. It is time DNA uses to seek perfection of life against changing environment. If time isn't realised in some form of realisation, then does it exist, did it exist to us before life and will it exist after our lives end? Is our short time alive our realised eternity? I believe it is. Things start and end, there is no reason why life and realisation should be any different.

When I hear the common saying 'but there must be something after death' I conclude that this is the conclusion of the needy human insecure brain. I want heaven etc to exist, but just because I want an outcome, it doesn't mean it will happen, no matter how strongly I may want to believe it. Under current operational policy guidelines, I won't qualify for entry, yet I know many good people who would. Billions of humans having the same view on this subject does not raise the probability that they are correct.

Life is about survival first and foremost. Animals, with the exception of humans and domesticated animals, spend their lives seeking nutrition, water, safety and shelter. Humankind has, in most societies and cases, removed the need to compete for these.

Food and water are readily available as the human ability to plan, within the neo-cortex, has developed farming and refining of food distribution and delivery points. Billions of humans live in various types of housing across the world, supported by societal structures that promote law and order while enhancing safety and security. Very little of human time within societies is spent on any or all of the survival essentials above; we now take them for granted.

Humans have an average of between 50-85 years of time to fill, depending on where they live, with a highly developed brain capacity which enables complex thought. The available 'downtime' away from survival 'essentials' has grown as civilisation and its characteristics have developed. Over the past century, that available time for leisure and complex thought has expanded more with

CHAPTER 5

the advent of better nutrition foods and supplements, increasing brain capacity yet further.

Since the launch of the internet, we have a bottomless reservoir of information, education, and learning techniques just a touch away. One of the by-products of the capacity of the human neo-cortex appears to be the increased capacity for emotion and with it, insecurity, superstition and therefore vulnerability. A need for things other than food, water, shelter, physical safety and reproduction. Unique human need.

It is as if the legitimate survival needs of belonging, safety and emotional security has manifested within the human brain, to provide a further need for survival non-tangibles, such as religion, cult, extreme politics.

Religion especially, but in many parts of the world its younger bedfellow, politics linked to nationalism, in the sense of belonging to a country, seems to exacerbate the need to belong in the human mind, to an extreme level in some circumstances. The great human neo-cortex, the most prized and amazing outcome that 3.8 billion years of evolution has given humans, appears easily manipulated by external forces such as 'false knowledge', 'certainty about the uncertain' and acceptance that what it is communicated by others to be correct, without question or need for evidence.

Amazingly complex and advanced as the human brain and its relatively huge neo-cortex is, it is easily manipulated.

Religious belief, but not exclusively, gives us obvious examples of this. In most cases, a person born in South America will grow up as a Catholic, considering the Pope to be God's representative on Earth, and viewing Jesus Christ as the most significant figure in history.

A person born in the Middle East typically identifies as Muslim and considers Muhammad the most influential figure in history. A child born to atheist parents in an agnostic or atheist environment is highly unlikely to believe in a human-described God. The common denominator, children overwhelmingly believe what parents and other adults tell them, and with no access to alternative information, are likely to continue to believe it for the rest of their lives.

Ancient Greek, Roman, Inca etc all had their own religious views, their children following the elders' lead. Why? Because the brain of the human beings who were informed of these 'facts', firstly had the capacity to understand, secondly had no reason to doubt what their elders, betters and system of education told them. The religions that were followed, started in one person's mind, or a number sharing ideas, and spread with many additions along the way, as they suited those who made or approved them.

A society requires both structure and adherence to that structure to operate effectively. The higher the degree of conformity the more effective, from the point of view of those who introduced them, a society functions, against its book of rules. This in turn gives security to those who hold

CHAPTER 5

power. Security and therefore maintenance of the status quo through conformity enables those with power to satisfy their own needs of wealth and therefore all basic survival needs are guaranteed well into the future. Survival, security, shelter, food and procreation accompany this.

Human need, through its highly developed brain, goes much further. A need for accumulation of wealth, planned increases in security and a buffer zone against enforced change usually accompany this human dominance through compliance of the recessive masses. Hierarchies developed, armies raised, tax systems introduced, laws created around the needs of those creating them.

Democracy has altered this dynamic yet maintained a level of corruption, of some of those in power. Politics has taken on the mantle in democracies, but a large part of the Earth still has countries which form laws almost purely based on religious teaching or extreme political doctrines.

Whether the reader believes these interpretations to be correct or not is maybe not the important question here. It is far more important to view the consequences of the result of how civilisation and society brought us to where we are. History informs us that countless billions of people have lived and died under various totalitarian regimes, whether they involve tribal, religious, or political elements.

A substantial percentage of these most likely believed at the end of their lives that they had lived fulfilled, happy, well-nourished lives, in safety and security. The primary reason for this seems to be that they were unaware of any alternative lifestyle. That is not necessarily saying that if they had wanted freedom and revolted, their lives would have been in any way improved. Is life more meaningful in a well-nourished autocracy or in a poverty-stricken democracy? Survival is still the main driver for the vast majority of humans – religions and politics have been added only recently.

Many revolutions lead to significant changes in societal structure, and they often occur alongside considerable suffering, frequently manifested in conflicts that result in loss of life. In the aftermath when society again strives for structure and laws to suit the new, it doesn't necessarily result in upgraded leadership. This is likely to, at least temporarily, disrupt food supply and other givens that threaten the basic characteristics of survival.

Revolutions in Russia and changes in Zimbabwe during the 20th century did not improve the living standards of the general population and, in many instances, worsened their conditions. The results showed parallels with George Orwell's novel Animal Farm, in which those in power prioritised their own interests above all else, harming the very people they claimed to serve. They merely replaced those who had led their countries on the wrong path, along another wrong path.

Democracies around the world applauded the end of apartheid in Rhodesia (Zimbabwe), apartheid was wrong. Yet when we compare basic survival needs of Zimbabwean citizens, for those from the minority factions within Zimbabwe, it could be argued that many were more disadvantaged under the Mugabe dictatorship than under apartheid rule.

Many black individuals faced the dichotomy of living within a poorer functioning economy, with many hardships and the illusion of freedom rather than under a regime which saw them as racially inferior, but where food was easier to acquire. Many with needy children, poor healthcare and rampant inflation may have viewed the time of apartheid rule with nostalgia. This doesn't make apartheid right in any form; however, it does demonstrate the superiority of survival needs over concepts such as freedom. Survival is embedded within DNA, freedom merely a concept of the human brain.

Both countries, however, proclaim the changes as magnificent epochs in their history. Joe Stalin, the Soviet Dictator was responsible for 20 million deaths, around fifteen million fewer than Adolf Hitler and possibly 20 million fewer than Chairman Mao in China; history records Stalin and Mao as heroes. Yet both murdered more of their own citizens than did despot Adolf Hitler.

Stalin ordered his party Commissars to shoot retreating soldiers making himself, rather than Hitler, responsible for an unknown yet significant number of Russian military deaths. This to add to the 700,000 he had executed in the purges of 1937. Not to forget the 3-5 million Ukrainian civilians starved to death between 1930 and 1933. Named 'Holodomor' this period is still remembered by the Ukrainian people as its darkest time, more so than World War II and the post-war retributions of Stalin for understandably supporting Nazi Germany.

Wars are usually a temporary occurrence, justified by the victor as worthwhile, the main factor usually being that it has maintained power or given power to a new leadership who will exact their ideas, will and laws on the human population.

Imagine that you are born in Britain or France in 1340 with a then life expectancy of 54 years. You are born at the start of the hundred years war, duration 116 years, and will die 62 years before peace. Children born to parents at an average age of 18, will know war all their lives as will their children and their children's children. It will be great, great-grandchildren aged well over 40 that will see peace between France and England.

War sustained through generations, higher taxes, enforced conscription and more draconian laws inevitably brought about by war, all tolerated without full revolution. There was civil unrest but no civil war. Why was war so accepted and tolerated, to the extent it was, for so long? A number of factors but high amongst them the human need to belong, and the safety and security engendered by that primal need to belong. The fear of change, the consequences of trying to bring about change, and the uncertainty of what

CHAPTER 5

would replace the existing environment, also likely to be at the forefront of citizens' thoughts.

The answer also lies in that, the war was intermittent, not a war as we view them today. There was respite and ironically the longer the war lasted the more it engendered and forged a sense of nationhood and rivalry. A rivalry with France which exists, to a lesser extent, within and between our contemporary nations.

So where is the balance tipped, when does the complex thought of the neo-cortex say enough is enough, I will override the fear of consequences? The need to belong appears never to diminish in humans; it has likely been a key driver of tribal, religious or political growth since humans existed. Evidence appears to show the need to belong can, however, switch allegiance. A driver for this in the past appears to be the threat to the basic survival needs.

In 1917 food shortages and war were the main drivers that led to substantial change in Russia. The French Revolution took place for similar reasons, related to food shortages; those who took part, risked all. The need to eat and survive outweighing the belonging to the state and the risk of death prescribed by the state.

In the 1920s and early 1930s the Nazis took advantage of the world financial crash and its exacerbated effect, together with war reparations imposed by the victors on the German economy and people following World War I.

Imagine you lived in pre-Nazi Germany, are married with three children, your children are without shoes, transportation, have limited education plus poor healthcare. You have no work and are reliant on a day's intermittent casual work and /or family handouts. A scenario not uncommon at that time. You have to spend any money immediately with hyperinflation making your pay rate on any given day, almost valueless, by the next day.

The survival factors of food, shelter and safety are running high on your agenda and running way past what political party is right for your country, in the longer term. Many millions more are voting for what is best for them, immediately.

If you had informed a reasonable man with children who had little hope in life, that the political party they were about to vote for, would unjustly murder millions of innocent people in years to come, how much difference would it have made to their voting intentions? The 'here and now' versus the unknown presumed future, survival without struggle outweighing the lives of those unknown to you.

It may be easy for us to judge the German people with the benefit of hindsight and from history. We did not lack hope, after living through more than a decade of poverty. We can though apportion blame to the individuals who exploited this human suffering, the Nazis, and possibly those who imposed the reparations that enabled the

CHAPTER 5

environment for National Socialism to propagate and thrive. After 1945, victorious western governments learned lessons, by recognising the consequences of imposing unsustainable reparations on Germany.

The western allies invested in West German reconstruction whereas Russia chose to impose reparations and dismantled East Germany manufacturing infrastructure, shipping it East. A decision the East German economy and people still bears the consequences of today.

Many people concur that the actions of the Nazis and the German state from 1933 to 1945 mark one of the lowest points of humanity's cruelty against each other in modern history. This is reflected history by us, the victors, and with good reason.

But, transporting ourselves back to mid-1930s Germany reveals a population largely ignorant of their government's dark side, flourishing with their basic survival needs met. Food, employment supplied in full, communities exercising and holidaying together. A higher sense of belonging and therefore security brought about not only by activities and lifestyle but strongly supplemented by propaganda. A number of survival and thriving boxes ticked by state initiatives, national pride restored, following post-World War I national humiliation.

Before the Nazis, propaganda existed in its basic forms, but they elevated it to an art form that fostered a strong sense of belonging among the German people while excluding those labelled as 'undesirables' for opposing their ideals, whether through actions, or simply by being different. In the modern era, propaganda has been refined into marketing, a force we meet daily. When you repeatedly or convincingly assert that something is good or bad, and no alternative view is allowed to be expressed, a human brain will likely accept that belief, whether it's true or not. Whether politics, the best washing up liquid or religion.

The human brain often finds it easier to comply rather than to question or resist, particularly when facing serious consequences for non-compliance.

As the Nazi state began to deteriorate following its losses at Stalingrad, in North Africa, and across Europe, the propaganda machine ramped up its efforts to pinpoint enemies of the state, outline the repercussions of opposing authority, and highlight the risks associated with a potential defeat. At this stage tens of thousands of Germans lost their lives. The State preyed on those primal survival factors, security and safety making sure this engendered fear to such levels that few Germans were prepared to publicly challenge the inevitable collapse and defeat of the State.

Remembering the savagery displayed by German armies moving through Russia, the German propaganda machine informed the German public to expect rape, theft, beatings and murder when the Russians arrived. In many cases the Russians met expectations. The distant sound of gunfire announced the Red Army's advance.

CHAPTER 5

When Germans recognised that defeat and submission to Russia was unavoidable, tens of thousands ended their own lives, with many also taking the lives of young family members beforehand.

The Nazis were a desperate story of human tragedy, both within, and outside Germany. The tragedy of Allied soldiers is rightly remembered every year, and on other significant dates, the tragedy of Axis soldiers less so, but let us not forget, many had little choice but to fight and die for a war many didn't believe in, or were fooled by propaganda into believing that war was just.

Daniel Goleman makes the following points. To better grasp the potent hold of the emotions on the thinking mind and why feeling and reason are so readily at war, consider how the brain evolved. Human brains, with their few pounds or so of cells and neural juices, are about triple the size of our nearest cousins in evolution, the non-human primates. Over millions of years of evolution, the brain has grown from the bottom up, with its higher centres developing as elaborations of lower, more ancient parts.

The primitive brain cannot be said to think or learn; rather, it is a set of pre-programmed regulators that keep the body running as it should and reacting in a way that ensures survival. The DNA programmed brain reigned supreme when the Earth was dominated by reptiles.

From the most primitive brain, emerged the emotional centres. Millions of years later in evolution, from these emotional areas evolved the thinking brain with its jewel, the neo-cortex with its great bulb of convoluted tissues that make up top layers. The fact that the thinking brain grew from the emotional brain reveals much about the relationship of thought to feeling – there was an emotional brain long before there was a rational one.

I believe the points Goleman is making is that human decisions are made from a mix of feelings and reason. If human feelings and emotions can be stirred and confused in the way the Nazis achieved through propaganda (brainwashing), then decisions can be on a mix of emotions and more primal reasoning. Instinct rather than rationality, dominating decision making.

The human neo-cortex possesses exceptional qualities, but it often tends to accept one-sided information without questioning it. The consequences of challenge can be too high – in Nazi Germany, Stalinist Russia, Mao's China, most likely death either through execution or death camps. The planning capacity of the human neo-cortex to link actions to a sequence of consequences, will prove a deterrent in most cases, as the need for survival trumps other thoughts of possible actions.

One could say the same about religion. Under the Catholic church as late as four centuries ago, it was heresy, punishable by death, to disagree with the church and its teachings, as Galileo found out, narrowly escaping death when he denounced his own, and Copernicus's scientifically evidenced discoveries. His daughter

CHAPTER 5

persuaded him to renounce what he knew to be true. The church reacting much like a child when told Santa Claus doesn't exist, with death threats for saying it.

Nicolaus Copernicus, a Polish priest and mathematician from the sixteenth century, is commonly known as the founder of modern astronomy. The credit goes to him because he was the first to conclude that the planets and the Sun did not revolve around the Earth. Yet to understand the contribution of Copernicus, it is important to consider the religious and cultural implications of scientific discovery in his time.

Copernicus, aware that his findings contradicted the Bible, feared the disdain of both the public and the church; he therefore devoted several years to privately refining and expanding his commentary on his work. The resultant work translated was entitled 'On the Revolutions of Heavenly Spheres' which was completed in 1530 but withheld from publication for thirteen years, until the year of his death in 1543.

When released, the work was attacked by protestant theologians who believed that the premise of heliocentric universe, effectively saying the Earth orbits the Sun and is not the centre of creation, was unbiblical. Because of this clerical opposition, and incredulity at the prospect of a non-geocentric universe, between 1543 and 1600, fewer than a dozen scientists embraced Copernican views. We can now bear witness to the ignorance of the prevailing religious view of that time.

"Religion is stubbornly ignorant, superstitious, impermeable to rational argument, lazy, narrow, shallow, and prejudiced". A C Grayling

Richard Dawkins makes a case against religion, and his penetrating analysis of how the virus of (religious) faith replicates itself, are among the most important of his contributions to debate in the philosophy and sociology of science.

Today's world has transformed significantly due to the emergence of a new information age: the world wide web. A human with access can find out information in seconds. When this information is factual, for example, the chemical composition of water, the date of the Battle of Hastings, the gold medal winner in any given sport at the 1984 Olympic games, it is not controversial.

However, just like religion and politics, humans providing opinions, or worse stating facts that simply are not true have been unleashed. What it has done is brought 'believers' together across the globe. Whether flat-earthers, anti-vaccinators, global warming protestors, pro or anti-religious groups, they now have a much stronger sense of belonging, where previously, they would have been more likely to have kept quiet about their views, for fear of being alienated or ostracised within their communities and losing the opportunity to belong.

I myself demonstrate this need to belong in the company of friends. I have friends on many levels as many people

CHAPTER 5

do in modern society. Having had a basic working-class upbringing like myself, a substantial percentage of my friends will discuss areas that affect their lives, sport, employment, TV, news coverage, politics, relationships, family and friends. I therefore stick to the safety of those areas of conversation, which I enjoy also. I feel comfortable and belong.

However, a part of me longs to discuss the discoveries of James Webb and Hubble telescopes, how Nazi Germany rose, fell and should never be repeated, how Edward III became arguably England's greatest King. How, in my view, modern religion has its roots in humans' domination of humans rather than anything to do with God or faith which appear convenient camouflage for its main purpose. If I talked about these subjects, my friends would probably start feeling disconnected since they wouldn't be very interested in those topics, and this might weaken our bond.

My human neo-cortex considers action and consequence and concluding one is not worth the other. Countless humans have pondered this before me. The need to belong is far greater in this case, than the need to express myself. The religious reader might find this perspective unusual. How can I make provision for friends, but not make provision for a human conceptualised god? My answer, I believe my friends exist.

The negative experience of scientific giants leads me to wonder how many potentially great scientists and discoverers 'we never had' because of the constraints placed on them by religion, especially through guilt and consequence. How many researched science to discover conflict with religious belief and therefore discontinued this work because of fear of consequences for themselves and their families?

Those whose survival instincts and need to belong killed off the sharing of their knowledge and contradicting religious text. Courageous individuals like Galileo, Giordano Bruno, and Darwin risked everything to defend their intellectual insights and the truths identified by their neo-cortex based on corroborated knowledge. Those people brave enough to overrule the part of the brain that tried to choose safety and shelter because of fear of being vilified, ostracised or even put to death, in favour of expressing truth.

Many, including myself, would not have been brave enough to risk upsetting the status quo. It is why I have so much admiration for these pioneers. I would willingly step between a violent or dangerous person or an animal and a friend or family member to protect them at risk to myself, but in comparison to the actions of Darwin or Galileo I would not consider myself courageous.

Those who sheltered people targeted by the Nazis, Stalinists or Maoists must have had similarities to heroes described above. Those who put the complex rational thought generated by upbringing, education, experience etc that led them, and their considered brain, driven by the neo-cortex and made complex decisions, in favour of truth. This, rather than being led by the more primal instincts of survival and self-preservation in other areas

CHAPTER 5

of the brain, dominant in the vast majority of human brains, including my own.

Humans have enabled themselves to act against instinct by over-ruling it within life learning and belief. Staying to fight unwinnable fights or self-destruction for a cause unrelated to preservation of offspring or procreation, are not actions that would be entertained in the animal kingdom. Animals will abide more closely to the blueprint of their DNA.

Insects through to fish, amphibians, reptiles, birds and even mammals, will live almost identical lives to others of their own species, strongly influenced by pre-programmed DNA. The simplest form of behaviour in animals is reflex, an innate stimulus response reaction which occurs by way of nerves that directly connect sensory input systems with muscles.

DNA programming usually includes survival staples such as seeking food, shelter and procreation. A wealthy human in New York City will live a very different life to a human in the rainforests of the Amazon. Both have essential survival needs functions of eating, sleeping, shelter and security, defecating and reproducing, but there the similarities end.

One could argue that until the arrival of modern humans, all animals and plants lived within strict behavioural confines dictated by DNA in their quest for survival. This could be in some degree by experience and accumulated learning through memory.

'Some plants actually anticipate sunrise from 'memory' and even when deprived of solar signals retain this information for several days'. Daniel Chamovitz

LeDoux tells us that plants and organisms can be said to be cognitive creatures. Some do this with information processing. Since all behaviour involves information processing, all behaviour would, under this theory, involve cognition. Others take a different tack, defining cognition as the adaptive regulation of states and interactions by an agent with respect to the consequences of the agent's own viability. With both such moves, all organisms are cognitive creatures.

LeDoux also informs us that plants anticipate sunrise from 'memory' and even when deprived of solar signals retain this information for several days. In 'Brilliant Green', Stefano Mancuso and Alessandra Viola argue that plants possess not only the senses of sight, touch, smell and hearing, but more than a dozen other capacities that humans lack, including the ability to detect minerals, moisture, magnetic signals, and gravitational pull.

HUMAN BRAIN 'SUPERIORITY'

We, as humans, often greatly overestimate our own significance due to our thinking and reasoning faculties. God loves me, 'he' made me in 'his' image, how very arrogant when we are just a small link in an evolutionary

CHAPTER 5

chain that has led over 3.8 billion years from single cell to highly complex intelligent life. Without every necessary link in that chain, we would not exist. Humankind is not, 'by a long way', the finished article. The difference between the most and least intelligent of us tells us that.

A person with an IQ in the 180s meeting another with an IQ considerably under 100 and conversing would demonstrate us as two different types of humans. We are still evolving to something else, which is quite a relief – what exists currently can't be a finished article, let us hope. When a high IQ accompanies highly developed emotional intelligence, this increases the scale and gap between humans. Young Hoon Kim of South Korea has a tested IQ of 276. Many individuals have high IQs but relatively low emotional intelligence; this can make them awkward in communicating and diminish empathy, making them appear 'odd' and poor at social integration. Intelligent but without key skills to live a fulfilled life.

David Goleman tells us, in a sense we have two brains, two minds and two different kinds of intelligence: rational and emotional. Our success in life relies on both IQ and emotional intelligence; neither alone is sufficient. Indeed, intellect cannot work at its best without emotional intelligence.

Hopefully the future versions of humans will not have excessive greed, the need to dominate and kill others in the name of religion, politics or country and these will not be a prominent theme in the next species of post-human DNA. Other human characteristics that bring down humans should not bring down what the human race will become, if still around, in 100,000 or even a million years, let's hope.

Every form of life on Earth competes to evolve in response to a changing environment. Each stage of evolvement is likely to inform pre-birth animals, through DNA modifications albeit at a very slow pace. For example, humans have little need for an appendix connected to their bowel; this is a remnant of when our ancestors ate amounts of raw vegetation and cellulose, but humans still have one which grows marginally smaller with each generation. This demonstrates that physical evolution can be guided by thousands of generations rather than just a few. The human brain is defying that trend, achieving rapid advances in just a few generations.

CONCEPTUALISING GOD

Imagine for a moment there is no God. This would mean that God on Earth exists only in the minds of those animals who can conceive of a god(s). Primates that have a neo-cortex functional enough to rationalise and self-describe a god, from information fed to it, by others with similar brain capacity. In a word, humans.

If a chimpanzee, or any other animal were to have thoughts about a supreme invisible power, no other

CHAPTER 5

chimps would be party to that belief – an inability to communicate and articulate to that degree of detail would disable the process. We can therefore conclude that belief in God, and spreading of 'his' word is reliant on communication to a level only humans possess, and a level we didn't always possess.

The need is for one brain to conceptualise, and ability to communicate the concept and that those communicated to can re-conceptualise using the information given. Humans may have very different images, of the same god in their minds, some with grey beards, others as an invisible breeze that pervades everything and everyone, plus other variations. My image of God was, for many years, taken from a book I read at six years old; it never changed.

We can be certain that no single cell creature has the ability to conceive a supreme being. Continue up to the great apes widely accepted as having the next level of brain capacity to a human of all land-based animals, it is also highly doubtful that the brain of a chimp or gorilla has sufficient capacity to conceive a god and the purpose it fulfils. A chimp would no more understand the concept of God as the internal workings of a combustion engine.

It appears that humans of 10,000 years ago didn't have the cognitive capacity to describe and believe in a god, at least not the Christian God humans conceive of today. That would mean that well over 96% of the time Homo sapiens have existed, they did not have the capacity to conceptualise a god as described by modern theologians.

The brains of animals appear to be hard wired to survival and reproduction only, as we humans once were. No matter how many times I communicate the concept of a god to a Chimp they are unlikely to follow it as it would not be significant to lifestyle and survival. Teach a Chimp how to use a tool for food collection advantage, and it will carry it in its brain, for life.

We must conclude that the development of the neo-cortex and enhanced brain capacity in humans, allowed for the conception of deities, as no mental imagery of gods was likely to exist before then. That, on Earth, God didn't exist in concept. Therefore, it is reasonable to conclude, if all humans died tomorrow God would no longer exist as a concept, just as the concept of God is unlikely to have existed for the first 3.8 billion years of life on Earth. It appears that God has been a temporary resident of Earth, and beyond, in the minds of those who conceive and believe.

As it is highly likely God didn't exist in the minds of Earth dwellers for almost the totality of life on Earth, something happened to change that. It is likely that improved nutrition through farming of crops and animals boosted human brain capacity through a newly found concentration of nutrition, which accelerated the development of the brain's capacity for complex thought. Survival needs became less of an issue; higher reasoning created new, less basic needs.

As discussed earlier, modern and improved food supply also gave humans more leisure time. The advent of

CHAPTER 5

art and learning, evidence of more leisure time and considered action, also longer residency in a specific area, unlike the nomadic 'follow the food supply' life of hunter-gatherers of the previous seven million years. It is believed that farming in a sufficient form to create this change to society, has existed for less than 10,000 years.

Ten thousand years as a percentage of 3.8 billion is less than 0.00026%. This suggests that God has not been present in the minds of Earth dwellers for more than 99.999% of the existence of life on Earth, as brain capacity was highly likely to be insufficient to conceive god(s). I doubt that a god who requires worship, would create such an outcome. If calculated from the earliest forms of primitive human religious symbolism, say 200,000 years, religion would have existed 0.005% of the time life has populated Earth. Conversely, it didn't exist for 99.99% of the time life on Earth has existed.

If it were proven that humankind could conceive of modern omnipotent gods 200,000 years ago, which appears unlikely, this would only make a difference to 0.009% of the time life has existed on Earth.

This is an excellent reason for some religious institutions to maintain apparent dogma regarding how long Earth has existed, how it was created, and when life was created. The sequence of events set out by modern science appears to contradict most religious interpretations. The undeniable fact that humankind is just part of a continued evolution and biological journey from single cell creature to more and more complex creatures.

This is at loggerheads with the religious concept that God created humankind in 'his' own image and the strong suggestion that we are the 'finished product'. I look around me in a media-rich world of news coverage to conclude, 'I hope to God that isn't the case'.

"I cannot believe in a God who wants to be praised all the time" Friedrich Nietzsche.

Throughout history, powerful people such as kings, emperors, dictators etc have wallowed in being praised and even worshipped, by their subjects; it stimulated the pleasure zones of their human brains. It was a confirmation of power and acceptance, God-like. They presumably assumed that any god that existed would require the same. A typical human brain conclusion. Like Nietzsche, I can't conceive or believe in a god that requires worshipping – it appears a very ungod like thing to expect.

I met a retired gay priest in a Liverpool art gallery. He had hidden the fact he was gay until he retired, his sexuality would have been viewed as against God for all of his time as a priest. We had a short conversation about religion. He, like most religious people, defended his beliefs by saying 'if there isn't a God then what's it all about, it could not have been an accident'. Why did he believe his version of events was correct?

CHAPTER 5

Because he had been exposed to indoctrination from being a child, maybe? Or the true explanation was beyond his reasoning capability, as it is with many people accepting a wrong answer, in preference to no answer. The need to have a believable, potentially wrong answer, appears to be greater than the absence of an answer.

He could ask the question because he had the brain capacity to do so. Yet he didn't have the knowledge to answer the question, so like many substituted with a plausible answer. If I had met a chimp, a dolphin or a penguin, it is unlikely I would have had a similar communication, if they had the ability to converse. Therefore, the conclusion I arrive at is, that the first time the priest's question was asked, was within the last 6,000 to 10,000 years, when humans had the capacity to ask it.

Then the reply to the priest's question is 'why must there be a reason, because you want to satisfy an unknown, your brain insists on complete and logical information on this subject and a simple answer doesn't have to exist because your brain demands one?'. There are many things humans don't know, yet we remain content, but the existence of an invisible being isn't included in that thought process – it appears, we must have answers.

Consistent religious commentary I have been subjected to is 'who was responsible for the Big Bang if there isn't a God'? A further example of the needy, intelligent human brain having to have explanations that it can understand and easily accept. Why this explanation, as opposed to the many there may possibly be, humans may not yet have the science or brain capacity to understand.

Some time ago, the church struggled to grasp Copernican or Galilean science, which clearly demonstrated that the Earth orbited the Sun. Out of ignorance, it threatened to punish with death, those who disagreed with its perspective. Clearly a wrong perspective.

BEFORE THE BEGINNING

What existed prior to the formation of the Universe? Empty space? Scientists believe that Space before the start of the Universe was not empty, but possibly full of charged quantum particles, which could leap in and out of existence. Possibly multidimensional.

Existence as perceived by human minds is something that has to be recordable by at least one of the human senses, unless it's a god, ghost or some other form of the supernatural which appear to have dispensation within the human mind, and doesn't need existence to be proven.

Why can't something exist which isn't recordable within the limited human scope? Dark energy and dark matter are known to exist because modern humans have added the dimension of the 'laws of physics'. They exist but not to the five human senses, they can't be seen, heard, touched, tasted or smelled, yet they do exist. Many 'invisible' forces exist,

CHAPTER 5

for example between 400 billion and 100 trillion neutrinos pass through our bodies every second. Neutrinos, incredibly small particles, with the mass of thousands of times less than an electron, created in dying stars, blasting through space at almost the speed of light and almost undetectable.

For 300,000 years, Homo sapiens were unaware of the workings of space to the extent compared with how our contemporary knowledge has developed. We are far from knowing the full story but the acceleration of knowledge is impressive. The shortcomings of the human brain may be addressed over time through evolution and the higher intelligence that should accompany it.

In the meantime, humans are likely to replace lack of knowledge with superstition to explain the, as yet, unexplainable. Using a god 'who has always existed' to explain how the Universe and time began. This appears to be the modern equivalent of how the Catholic Church treated Galileo, because of lack of understanding, and therefore ignorance. It is the insistence to project that unconfirmed information, as knowledge.

Scientists admit their limitations, when they reach the limit of data or supported theory, that should not detract from their incredible discoveries and advances so far. It should not lead religious scholars to then fill any gaps in knowledge with ancient text or religious theory.

The church has faith in all its conclusions, at least the ones haven't been disproven, to date. The geocentric model clung to by the Christian church for 350 years after the death of Galileo has been discredited. In 1992 the Vatican admitted Galileo was correct when the rest of civilisation had known it for hundreds of years.

The human brain seems to believe that every destiny-based question has a clear explanation. That isn't the case at this stage of human brain development, in the same way as interstellar travel isn't possible, yet, and may never be. As human intelligence advances and we hopefully develop into a yet more intelligent creature, more answers to more questions will be forthcoming as will more questions that don't yet have answers.

In summary, humans are intelligent when compared with other animals now, yet in future more intelligent species will exist, probably evolved from humans, unless we are extinct before that time. They will not have the answers to all questions, but will have answers to many more than humans today.

This human 'adding of an answer' appears to satisfy the human brain's need for complete information in this sphere, whether it can be evidenced or not. This appears typical of the later development of the human brain and satisfying its need for explaining the unexplainable, with religion or other superstitious infills.

The insecurity surrounding what is 'believed' has meant that religious doctrine has needed protection for thousands of years – failure to believe or comply was met

CHAPTER 5

with death and the speedy destroying of resistance, by those with the most to lose, to prevent it being touted as a scam, or replaced by an alternative.

The killing of those found guilty of witchcraft was not purely about punishing perceived evil, but the sending of a message to all regarding the fate, which was likely, if individuals defied or rejected religion.

The further back in the history of humankind, the less capacity there was for complex, rationalised thought. Six thousand to 10,000 years could possibly be a wrong number – this is not an exact science. Over 10,000 years ago the human brain capacity and cognition was diminished compared with modern humans, until relatively recently within Homo sapiens' evolutionary journey. Flynn's research in a later chapter gives context to rapidly growing human intelligence over just 90 years, when given the right conditions.

We can see clear evidence of the cognitive and reasoning advances of humans over the last 1000, 200, but especially 30 years, particularly in westernised culture where education has progressed and standards raised, from a privilege of the wealthy elite, to a 'Right' of most citizens. My children tackling schoolwork at a level I had, but two years earlier, this is within a separation of 30 years, a clear sign that challenges and standards within education are rising to meet developing minds.

Huge developments in our ancestors' brain power have taken place over seven million years, after our ancestors split from the ancestors of chimpanzees to live lives with different challenges. Natural selection rose to the challenge of evolving humans, both physically and mentally, to adapt to new environmental demands. The brain power of Homo sapiens seems to have accelerated significantly since the cognitive revolution began 70,000 years ago, surpassing the advancements made over the previous seven million years. This sudden increase in brain capacity appears to persist and even accelerate, in modern times.

The last 10,000 years being a further significant advance on the previous 60,000, including the start of civilisations as we recognise them today. In the last 200 years, since the industrial revolution, humans have dramatically transformed the Earth, making advancements in cities, travel, mechanisation, and communication that surpass those of the previous 10,000 years.

The advancements over the past 30 to 40 years are yet more remarkable. My father died 43 years ago; he would hardly recognise our hi-tech world today. Advances made by the human brain appear to continue at increasing pace.

At birth, humans are completely helpless; we take about a year to learn how to walk, an additional two years to fully express our thoughts through language, and many more years before we can fully care for ourselves. We completely rely on those around us to survive. In contrast to many other mammals, such as dolphins that are born

CHAPTER 5

as swimmers, giraffes can stand shortly after birth, and baby zebras gain the ability to run in just 45 minutes. Yet once humans meet full maturity, the advances from birth are huge, with self-sufficiency and reasoning powers well beyond any other creatures on Earth.

*

References, sources and recommended reading or research, Chapter 5

Consciousness – Stanford Encyclopaedia of Philosophy (Plato Stanford)

Describing and explaining consciousness – Bjørn Grind, Neuroscience of Consciousness, Oxford Academic

Emotional Intelligence – Danial Goleman

The Deep History of Ourselves – Joseph LeDoux, observations of human culture and the human brain

How does the Human Brain create consciousness and why? – Harriet Pike, Medical News Today

The facts and doubts about the beginning of the human life and personality – Asim Kirjak & Ana Tripato

Religion and Geography – C.Park, Lancaster University

Structure of Civilization – Politechnika Warszawska

Revolutions, Modernisation and Contemporary Civilisations – Edourd E Shults

Exploring the Politics of Chronic Poverty: From Representations to Politics of Justice – Sam Hickey & Sarah Bracking, Science Direct, World Development

CHAPTER 5

The role of economic instability in the Nazi rise to power - The Holocaust Explained, Wiener Holocaust Library

Visions of Community in Nazi Germany, social engineering and private lives - Oxford Academic

Religion and Politics - Oxford References

Robert Mugabe: from liberator to tyrant - Joseph Winter, Africa Editor, BBC News

Unravelling the human code: ae humans programmed and coded - Parth Khajgiwale, Indian Times

How do plants sense the world around them - John Innes Centre

Forget me Not: scientists pinpoint memory mechanisms in plants - University of Birmingham

Micronutrients and the evolution of the human brain - Hans K Biesalski, Science Direct

The evolutionary roles of nutrition, dietary selection and dietary quality in human brain size and encephalization - Springer Link

Why did Witch Trials happen? A brief history of Christian takeover of paganism - Lauren Bayliss, North West Byline

Frank Drake, Astronomer Obituary – The Guardian UK

How many people have ever existed on Earth – National institute for Corrections

The Disintegration and Death of Religions – Oxford Academic

Hundred Years War: consequences and effects – World History Encyclopaedia, Mark Cartwright

How Racist Rhodesia did it and Independent Zimbabwe is getting it wrong: comparing currency, and finance under sanctions – Developing Economics, Tinashenyamunda

Russia's Home Front 1914 to 1922 – The Economy, The University of Warwick, M Harrison

Who killed most Hitler, Stalin or Mao – Jagran Josh, Nikhil Batra

What came before the Big Ban – buffalo.edu

CHAPTER 6

Universe

—

We have covered how Earth formed and cooled to eventually produce life. How that life struggled for 300 million years to get a foothold and replicate sufficiently so that birth of life was greater than life that perished. Unicellular life then dominated a harsh, hot, unforgiving Earth for more than two billion years until conditions were sufficient for multicellular life to take hold and eventually thrive.

How life grew and expanded to present day, countless varieties of land-dwelling animals exist, this in spite of five extinctions which threatened to snuff out life on Earth. Then, in the very latest part of that journey, humankind arrived at the end of an almost everlasting chain of evolution which should continue for at least hundreds of millions of years.

With humans came larger brains; these brains would go on to make many discoveries and inventions making the world we know today. One of the greatest achievements has been the exploration of space, observed by the ancient man including the ancient Greeks who documented distances, sizes and other characteristics of space. Copernicus, Galileo, Newton, Einstein et al who contributed so magnificently to what we know today.

In the last 30 years Hubble and James Webb telescopes plus many other contributions to space exploration have opened up a new world outside the previously and relatively 'small', observable Universe. The true enormity, strength, and magnitude of what you are about to read is absolutely astonishing.

How far from here were you born? The answer depends on old you are! If you were to die on your 70th birthday you would have died nearly 800 trillion miles away from where you were born. Our Galaxy is travelling at around 1.3 million miles (2.1 million km) per hour through space.

For 4.6 billion years our planet has been spun in an orbit around its star, and the star in an orbit around the centre of our Galaxy, 'owned' by a supermassive black hole weighing in at 4.3-4.6 million times the mass of our Sun (solar masses). The scale of it all is mindboggling, yet almost irrelevant compared with the happenings elsewhere in the Universe.

Over the last 30 years, especially following the launch of the Hubble space telescope, researchers have made incredible discoveries about our Universe. Discoveries

CHAPTER 6

that Albert Einstein, Edwin Hubble, Galileo, Copernicus, Isaac Newton, Kepler, Halley and many others 'giants' of scientific discovery, would have 'given a limb' to have lived to have witnessed.

This scale exceeds everything we, as humans, could have possibly imagined. Possibly as many as 200 billion to 800 billion, or more, galaxies, with hundreds of billions of stars in each, some containing many trillions. Some higher estimates put the number of galaxies at around two trillion, with the possibility of an existing Metaverse, containing potentially countless Universes.

The estimated age of our Universe is around 13.8 billion years (13.7 billion plus according to scientists) or to put that into context 46,000 times the 300,000 years that Homo Sapiens have walked the Earth. The Universe is likely to exist many millions of times longer than Earth and our comparatively small-medium sized star and its fragile solar system.

Consider the following. We are observing the Universe at one moment in time. To observe stars and planets 100 light years away, is to see them as they were 100 years ago, a galaxy 12 billion years away, to see as it as it was 12 billion years ago.

The question, 'is there life on other planets', should be altered to 'has there been, is there now, or will there be in future, life on other planets?'. These are very different questions in terms of scale, the question not only includes the billions of trillions of stars and planets that are likely to exist today, plus those in larger numbers that have existed and died, but also those that will exist in future. The simple answer is yes, the numbers are too large for our story, or similar, not to have been, to exist now, and be repeated or have occurred many times in the future.

If our instrument technology was effective enough to view a planet in detail at a distance of 4.5 billion light years and it was molten, we would conclude that life could not exist on it. Yet if we were observing our own planet 4.5 billion years ago, we would arrive at the same conclusion: the Earth was molten.

Similarly, if we could observe a planet teeming with life at a similar distance, in the 4.5 billion years it had taken for the images to reach us, that planet and its solar system could be long since dead. Our Universe is dynamic, change is constant; it will remain so for at least 100 billion years and exist for many times longer. Trillions of stars observed in distant galaxies no longer exist, they will have died and pass the dust they became onto other forming stars, planets or vanished into a black hole.

The search for life in distant galaxies will likely be a 'long gone' or 'yet to come' experience. The Universe isn't old enough for us to find life in the most distant reaches from Earth, as we are observing areas where temperatures, shortly after the start of the Universe, were too high to sustain life. Stars must form and then die to create essential elements like carbon, oxygen, and iron, as well as other

CHAPTER 6

fundamental building blocks of life. Without stars having died, complex life could not exist. We are, indeed, stardust.

A high percentage of our Universe is 'yet to come'. Trillions times trillions of stars are yet to be formed, create planets; some will create life, then die, in the ongoing cycle of mass creation and mass destruction, that is the Universe. Many will have a life form, yet the definition of 'many' may be a fraction of a percent. A much smaller fraction of a percent may have complex life, possibly more advanced than humans, but that relatively small fraction may be trillions in number, such is the numerical scale of stars and planets in our Universe.

The search by humans for extra-terrestrial life is likely to concentrate on our own galaxy, more than 100,000 light years across, or Canis Major, a Dwarf Galaxy about 42,000 light years from the centre of our Universe, yet at 25,000 light years distant from Earth, the nearest part of Canis, it is closer to Earth than the centre of our own galaxy.

Andromeda, the Milky Way's nearest large galaxy, is 2.5 million light years away and will give us a much older story of any lifegiving planets discovered and has the extra degree of difficulty in observing through an additional 2.4 million light years of space than the furthest extremes of the Milky Way.

Most life discovered in our galaxy will be a story of the past because of the time lapse; 10,000 years ago, humankind was hunter-gatherers, the Earth, observed from space, would have looked roughly the same as it had a million years ago. A relatively short time later, the Earth has cities, civilisations, hi-tech communications and space travel. Human habitation can be viewed from space across swathes of the developed, illuminated Earth. It took less than 200 years, making our planet look very different when viewed from afar, especially in darkness, when the lights of humanity are clearly visible from space.

Science fiction and short-range space travel in the 1960s to 1990s, led us humans to believe that inter-stellar travel may be possible in future. It is, of course, possible that it could be, we can't underestimate the human brain and its capability. However, recent discoveries have helped us realise the scale and degree of complexity of what may be needed.

The closest star to Earth sits about 4.25 light-years away, which translates to roughly 25 trillion miles. This may or may not have a habitable planet for humans, but this is unlikely. Let's for a moment assume it has one. The hostility of the Universe has become clearer since the launch of Hubble – the amount of debris in our solar system remaining floating around in space following its formation. In addition, a huge, yet unknown amount of debris is floating around in space inside and outside our solar system.

Travelling at the exceptionally high speeds needed to reach another star within a human lifetime would carry

CHAPTER 6

a high risk of collision with space debris – even very small objects would destroy any craft which impacted with them at the speeds necessary to reach other solar systems within the life expectancy of humans.

If we then assume humans develop the ability to travel one tenth of the speed of light, 18,600 miles per second. The UK to Perth, Australia in approximately half a second. It would still take more than 42 years to travel to our nearest solar neighbour. This figure represents over fifty percent of the current average human life expectancy. Travelling such long distances at such high speeds would use more energy than humanity has ever generated.

It is highly unlikely then, that it is a trip we could make for various reasons. Wormholes and other conveyances are theorised by scientists, better brains than mine. However, I find it difficult to believe during the formation of the Universe and the formation of spacetime, that a 'time bridge' has been conveniently built between two planets habitable by humans. The much higher probability is that if wormholes exist, they connect empty space to empty space, or uninhabitable to uninhabitable solar systems, possibly thousands of light years away from habitable planets.

Unless something changes radically and is unforeseen at present, humankind is destined to play a relatively temporary part in the life of the Universe. Humans have limited time to develop the means for inter-stellar travel;

how long, we can't tell. The race is on against the timing of the next extinction, whenever that may happen, but it will happen.

Extinctions of many species of life on Earth will continue at intervals until the Earth itself is destroyed. The destruction of the Earth is an eventual certainty but well before that the environment that supports multicellular life will cease to be. And before that life will have many challenges to face, especially complex life. It will win some, but eventually lose; it can overcome many, yet only needs to lose one, as the dinosaurs discovered.

*

CHAPTER 6

References, sources and recommended reading or research, Chapter 6

How fast are you travelling when you are sitting still – nightsky.jpl.NASA.gov

When we look at the night sky we are looking back in time – BBC Night Sky Magazine

Proxima Centauri: Alpha Centauri, facts about the stars next door – Robert Lea, Space.com

How the Universe Works – Pioneer Productions / Discovery Channel, 8.5

Brains to Brilliance: The evolution of the mind towards cerebral thinking _ The Oxford Scientist

CHAPTER 7

Incredible Scale

Scientists are in general agreement that the Universe was created by a massive expansion from a single point of pure energy, a singularity, about 13.7 to 13.8 billion years ago. In short, matter formed, and gravity attracted matter together to increase mass; this helped the later formation of stars and planets. This is a simplified version of how the Universe then went on to create galaxies and therefore solar systems with their stars, planets and moons.

Einstein demonstrated that matter and energy are intertwined, and one can create the other. An example of matter creating energy is an atomic bomb, for example. The splitting of the atom matter releases tremendous energy; an example of the reverse is the 'Big Bang' where energy created matter. Atoms can be viewed as a reservoir of energy, left over from the Big Bang; matter the size of a human hand contains enough potential energy to destroy the Earth. Energy creating matter and vice versa is the basis of Einstein's formula, $E=MC^2$.

CHAPTER 7

'Big Bang' may be misleading as it is unlikely it was an explosion. The expansion of the Universe was more uniform, faster than the speed of light, more akin to inflation of the Universe. Imagine a lightbulb being switched on and how light travels, at the same speed in every direction, except the Universe is believed to be flat, rather than sphere shaped, therefore imagine the light spreading more like a ripple on a pond. Compare that to an explosion where different parts of matter are projected at different rates in different directions. The lightbulb analogy appears closest to the start of the Universe, although still not wholly accurate.

Although the Universe is 13.8 billion years old, it is now believed to be over 93 billion light years in diameter. Scientists believe it to have expanded at faster than the speed of light in its formation. The laws of physics insist that nothing in our Universe can travel faster than light. That law applies within our Universe, the Universe expanded within space that was not part of the Universe, a millisecond earlier, and therefore scientists believe that the Universe expanded faster than the speed of light; this explains why it expanded across 93 billion light years in under 14 billion years.

When the Hubble or James Webb telescopes look across our Universe and receive light from a few hundred million years after the 'Big Bang', they are looking at galaxies that are no longer within viewing distance. We are seeing them as they were 12-13 billion years ago. Much closer than they are now and less developed.

The very furthest galaxies could now be more than several times that distance away. The light leaving them now may never reach our planet, firstly because they may be too far away, but also because our solar system will have died, along with our planet, many of billion years before that light can reach us.

Galaxies are moving away from one another at incredible speed. In 100 billion years our Galaxy will probably be isolated with no other galaxies in view, but will by that time have benefited from dark matter applied gravity, pulling it closer to, and merging it with the Andromeda galaxy, in approximately two to four billion years. There is still a gravitational and a probable dark matter attraction which prevents Andromeda and the Milky Way moving further away from each other. The reverse is true as they are drawing closer to each other.

Although isolated from other galaxies, the joint Milky Way and Andromeda galaxy will form a larger galaxy, having also consumed nearby dwarf galaxies that have wandered into the gravitational influence of Andromeda, the Milky Way or the later combined super galaxy. Star production within the combined galaxy will likely have ceased in 100 billion years, and its stars will be mainly red and white dwarfs, low powered stars or once larger stars that are low on fuel.

The speed that galaxies are moving apart demonstrates the power of the Big Bang and how it deceived the laws of physics to produce the energy that made matter travel at speeds faster than light, over incredible distances. Matter did slow from its high velocity following the Big Bang; it would have been expected that their speed of travel would have eventually slowed more. However, scientists proved that five billion years ago galaxies speeded up. It believed with the assistance of an, as yet, mysterious force, called dark energy.

In 2011 the Nobel Prize in Physics was awarded to Adam Reiss, Brian Schmidt and Saul Perimutter for this discovery. An invisible source of energy, little understood by science, as I write, but known to exist, through observing the forces and matter within our Universe, Dark Energy.

To summarise the scale and power across our Universe as communicated within this book

THE UNIVERSE (CONTINUED)

Is believed to be 93 billion light years in diameter, source national radio astronomy observatory

Contains a minimum of 200 billion galaxies containing a minimum of 2×10^{22} stars (20,000,000,000,000,000,000,000). NASA estimates the number of stars could be a Septillion 1×10^{24}

Hosts massive galaxies with estimated star numbers in excess of 100 trillion, between 250 and 1000 times the mass of the Milky Way and over 100 times longer. Discovered as recently as 2022, the Alcyoneus galaxy, located approximately three billion light years from Earth, is over 16 million light years across. If our Milky Way was placed inside Alcyoneus, it would be similar in scale, to putting a small sofa into Wembley stadium. At three billion light years distance, the size was therefore measured as it was, three billion years ago, substantial time for it to have grown yet more or to have diminished in size.

Frank Drake estimated that planets similar to Earth, with life forms capable of sending signals into space, occur in orbit of around one in 20 million stars. Using the conservative estimate of stars above, that makes it probable, at least possible, that 1,000 trillion intelligent life-supporting planets exist at this moment in time throughout the Universe. This estimate is a current estimate and does not include those worlds which have lived and died, or those yet to come.

The Universe has had many times the stars it has currently, and over the next 100 billion years, it will possibly generate more stars than exist today, but in decreasing numbers compared with the earlier Universe. It is believed more stars may have been produced in 13.8 billion years than will be in the next hundred billion years; both of these totals are expected to be many times the stars that exists today. Future production of stars leads to a significant rise in the estimated number of Earth-like planets there are likely to be in future.

CHAPTER 7

Hosts black holes up to 66 billion times the mass of our Sun or around 20,000,000,000,000,000 (2x10^{16}) times the mass of Earth, the gravity is so strong in the vicinity of a black hole event horizon it can slow time to a standstill. This, from the small part of the Universe observed to date, it is therefore highly probable that much larger black holes exist and will exist in yet larger forms in future.

Hosts ultra and supermassive black holes combining events. These can produce a billion, billion, billion, billion times the energy of a supernova, an undecillion or 1x10^{36}. In the instant of joining, 'the joining' produces 100 million times the light produced by all the stars in Universe, in that same instant. For context, a large star turning supernova could damage or destroy Earth from tens of lights years distance. The level of energy released, although unimaginable, is only a fraction of the potential energy contained in the newly formed black hole.

Contains quasars and supernovae that can deliver more energy in a single second than our Sun will produce in its lifetime.

Medium and larger sized stars live for between a few million tens of billion years, which given the numbers that are likely to exist, this means that trillions of stars die each Earth day across the 93 billion light year of our Universe.

In the micro Universe 7000 trillion, trillion atoms make up the average human body, these make up around 30 trillion cells in the average human.

Yet smaller are neutrinos with a mass of thousands of times less than an electron orbiting the nucleus of an atom. Hundreds of billions can travel through a human body per second. Lighter yet is the photon, they have no mass, what we know as light. Light can bend around to pass, but neutrinos pass straight through.

HOW MANY STARS?

The question of how many stars have existed or will exist is a difficult one, as scientists don't know the full size or scale of the Universe. The two estimates of current stars used in this book so far are huge numbers, but the lower one is only 2% of the higher NASA estimate.

The estimated scale as mentioned earlier, is very large indeed. Possibly hundreds of billions of galaxies exist in our Universe with up to a trillion stars, some having tens of trillions and more, but usually hundreds of billions in an average sized galaxy. Our own Milky Way being a mid-sized galaxy with between 100-400 billion stars. The number of stars in the Universe could range from 20 billion, trillion to a septillion, 1×10^{24}, according to NASA, the higher number 50 times that of the lower estimate. The smaller number being difficult to imagine

CHAPTER 7

in scale, more than all the grains of sand on all of the beaches on the Earth.

STARDUST NOT GOLDEN UNTIL DEATH

The foundations of life on Earth were made in supernovae billions of years ago. A supernova is activated when a star approaches the end of its life. The metals and chemicals essential to life were formed in dying stars probably billions of years before our Sun and solar system formed. When a star goes 'supernova' its outer layers are blasted into space, releasing incredible levels of energy, and therefore potential matter. Remember $E=MC^2$, which confirms matter and energy are interchangeable.

A star generates energy through the internal process of nuclear fusion, where atoms come together under immense pressure, producing more energy than nuclear fission – fission involves splitting atoms. Without nuclear fusion life could not exist. This occurs for two primary reasons: first, life elements are generated, and second, our star's heat sustains life through nuclear fusion.

A newborn star holds a plentiful supply of hydrogen. Compressing hydrogen atoms under extreme pressure results in the fusion of two hydrogen atoms into one helium atom. This reaction will release high levels of energy, more than splitting the atom as happens in atomic bombs. The sun fuses 620 million metric tonnes of hydrogen every second.

At the beginning of the Universe, hydrogen was the only element present for at least one minute. The temperature rose very quickly; when it reached over a billion degrees Fahrenheit, hydrogen atoms were fused together. Scientists believe temperatures rose to 250 million, trillion, trillion degrees Fahrenheit, which resulted in incredible amounts of hydrogen fusion, making 25% of the Universe's matter, helium. The cycle of nuclear fusion was born and the foundation of all matter, creating stars, planets, moons and all life.

When helium becomes abundant it will act as the base for the formation of other elements. Helium atoms crushed together form beryllium and eventually carbon. Then the following is a simplified version of what will take place in the lifecycle of a star, to produce energy by nuclear fusion, the process is called Nucleosynthesis:

- hydrogen fused with hydrogen produces helium
- helium fused with helium produces beryllium
- beryllium fused with helium produces carbon
- carbon fused with helium produces oxygen
- oxygen fused with helium produces neon
- neon fused with helium produces magnesium
- magnesium fused with helium produces silicon
- silicon fused with helium produces sulphur
- sulphur fused with helium produces argon

CHAPTER 7

- argon fused with helium produces calcium and,
- calcium fused with helium produces titanium
- titanium fused with helium produces chromium and,
- chromium fused with helium produces iron

These, and other fusions – this list is exemplary and far from complete – are the building blocks of everything we know. All the elements of the periodic table heavier than hydrogen were created by nuclear fusion.

Iron is almost 56 times the weight of hydrogen and is therefore the result of numerous fusions. The solar process of creating heat over its lifecycle crushes and fuses light matter into heavy matter, releasing immense levels of energy, eventually leading to burnout and death of the star.

The list of fusions above and more varied ones don't take place in an evenly balanced way over time. Hydrogen burns almost throughout the life of a star and when the hydrogen is exhausted and heavier elements dominate, the star is near the start of its 'death throes'. The dying star is now in distinctive layers with heavier elements closer to the core and lighter elements nearer the surface, in order of mass. By this time the elements within the star are becoming too heavy and gravity is stronger when mass is greater and heavier. Energy diminishes and gravity increases. When the death of a star occurs, it is triggered in an instant.

The formation of excess heavy metals such as iron, signalled that the star was very close to dying, and supernova would disperse the accumulated mix of chemicals into space, to eventually, possibly be reused in another solar system cycle. The reason we exist.

In the process of exploding, when in supernova phase, heavier metals such as gold and uranium are created. The Universe had plenty of time from its inception to the birth of our Sun to form and destroy other stars and transport the remnant matter to our solar system, which culminated in the creation of our sun around five billion years ago.

Except for hydrogen and helium, all atoms of elements were created through astrophysical processes, such as supernovae, collisions of neutron stars and high energy particle collisions. The heaviest particle known to have been created is uranium-238, named because it is 238 times heavier than hydrogen. Around 98% of the Universe is made up of either hydrogen or helium, meaning only 2% is made of other elements. A plentiful supply of raw materials to produce elements in stars, yet to be created.

We live a macro Universe made up of galaxies and stars, driven by a micro Universe containing atoms, sub-atomic particles and other forces. A minimum of 20 billion, trillion, stars above (around) us, an average of 7000, trillion, trillion atoms within us, the scales at either end of the size spectrum are immense. Objects smaller than atoms, to objects up to 66 billion times the mass of our Sun have been proven to exist within our Universe.

CHAPTER 7

Our solar system includes the remnants of at least one long-deceased star, and likely from more. Days 3,5 and 6 of biblical creation cover this phenomenon; we can be sure it took longer. There will be no horsemen involved in its death around five billion years from now, just a red expanding hell, consuming planets as it grows. The last remaining life will be unicellular life forms buried miles beneath the Earth's surface and the surface temperature increases from hundreds to thousands of degrees Fahrenheit.

A future star yet to be formed, or stars and their planets, possibly even life giving, may share the metals and chemicals from our collapsed star and its surrounding planets and their contents, including the elements, once contained in human bodies. Regeneration and circularity are overwhelming themes across the Universe within galaxies; much of this is orchestrated at their centre, by black holes. A Supermassive or Ultra Massive black hole is believed to be located at the centre of all, or at least the vast majority of galaxies.

GALAXY SUPER POWER

Our own supermassive black hole, at the centre of the Milky Way, is known as Sagittarius A star; it has a diameter of approximately 27 million kilometres. Sag A star is a mere 26,000 light years away, approximately 152 quadrillion miles. Weighing in at 4.3-4.6 million solar masses, our Sun is one solar mass, Sagittarius A star is 'small' compared with ultra-massive black holes discovered in other galaxies. The smallest ultra-massive black hole, classified as over five billion solar masses, is more than 1000 times heavier than Sagittarius A star.

Although scientists are not sure how a supermassive black hole is formed, they are certain about smaller black holes being formed by the collapse of stars above a certain mass. A large star will become a black hole where smaller stars will go supernova and burn out to a white dwarf, much smaller and cooler than the original star.

Stars vary in size, and their lifespans differ accordingly. Larger stars usually have shorter lifespans because they burn through their fuel more quickly. However, even after their normal lifecycle, they still retain significant power. Large stars, which live shorter lives than small or medium sized stars, may collapse within a fraction of the time, yet larger stars are likely to form black holes and, in that form, likely to exist for hundreds of billions of years.

Before a star dies and goes supernova or implodes into a black hole it is locked in a battle. There is an ongoing fight for the existence of the star between gravity trying to crush the star and the energy produced by the star pushing outwards. In the end gravity is the victor, the elements within the star get heavier and heavier, making it harder to produce energy to push outwards and resist gravity.

Gravity compresses the star to a fraction of its original size, yet a substantial portion of its mass stays concentrated

CHAPTER 7

in a much smaller region, as the denser core maintains a powerful gravitational force. Scientists believe strong nuclear force to be 100 trillion, trillion, trillion times stronger than gravity, but when atoms can no longer be fused to produce strong nuclear (fusion) force, gravity wins.

The immense mass of a black hole exerts a powerful gravitational force that attracts matter toward it. One plausible theory for the formation of supermassive black holes suggests that they arise from the consumption of smaller black holes, which they attract through gravitational forces and grow in size cumulatively.

If accurate, the enormous mass of ultra-massive or supermassive black holes suggests that possibly hundreds of billions of stars have existed, died and been absorbed into black holes, since the Milky Way's formation. Possibly having life giving Planets and complex life, this being destroyed with the destruction of solar systems, the eventual destiny of life on Earth. The eventual destruction of solar systems being a fairly routine event across the solar system.

ASS Nova estimate that less than one per cent of stars are eight solar masses or over. At eight solar masses a star will burn for around 100 million years, stars with 10-15 million solar masses will burn for 10-20 million years and those of over 100 solar masses for as little as a few million years. This is the reason why complex life is unlikely to exist on planets orbiting larger stars. It took nearly a billion years for the Earth to produce sustainable unicellular life, three billion to produce complex life.

ROUTINE DEATH OF STARS

A caveat to what I am about to estimate is that stars are created in many sizes. Lifecycles from a few million to trillions of years. Rate of death obviously related to expected length of life but not exclusively. Rates of premature death could be related to how active or large a galaxy is, its size and number of black holes or proximity to other galaxies; two conjoining galaxies could raise attrition rates. Different galaxies will therefore have different levels of star attrition rates.

It is estimated that 0.8 per cent of stars are classified as large. Using the two earlier estimates of stars and Frank Drake's equation, I have made the following calculations.

$0.8 \times (2 \times 10^{22})$ stars being an average of around 15 solar masses and burning for 15 million years, I calculated how many large stars on average, die each Earth day, across the Universe.

2×10^{22} Stars $\times$ 0.8% = 1.6×10^{20} larger stars

Number of Earth days in 15 million years = 5,478,750,000 days, a proxy for the average life of a large star.

1.6×10^{20} divided by 5,478,750,000 = approximately 29,200,000,000 large star deaths per Earth day, an average of nearly 338,000 per Earth second Universe wide.

CHAPTER 7

If the NASA estimate of a septillion stars is used for this equation, then an average of 1.46 trillion large stars per day or nearly 17 million per second.

An unimaginable trail of destruction, each large star producing enough base material needed to create our solar system several times over, with all necessary elements for future life. An incredible level of recycling of matter.

For the purposes of avoiding complexity, I use our Sun and its mass as a proxy for all medium stars. Stars obviously come in sizes ranging from hundreds of solar masses to less than a tenth of a solar mass. William and Deborah Hillyard, Life History, define medium sized stars from 40% of a solar mass to 400% of a solar mass. They will have lifecycles from many times to a small percentage of the life of our Sun.

When using our Sun as a proxy for the size and length of life of a medium sized star, we must continue to remember, therefore, that the different lengths of star life can vary hugely. If the mean average is lower than the Sun's mass, then stars will live longer and the number of star deaths will be lower, and vice versa. But all stars will eventually die, making estimates over many tens of billions of years, for small medium to large medium stars, providing the overall estimate of stars has some accuracy. The life and death of stars is inevitable.

If ten per cent of stars in the Universe were a similar size to our Sun and live for around 10 billion years, then, using the lower estimate of stars:

$2 \times 10^{22} \times 10\% = 2 \times 10^{21}$ divided by 3,652,500,000,000, number of days in 10 billion years, an average of nearly 550 million medium sized star deaths per day or over 6,300 per second.

If the NASA estimate of a septillion stars 1×10^{24} is used in this equation, the average deaths of medium sized stars, at over 27 billion medium-sized star deaths per day, or over 316,000 per second.

Using NASA's higher figure of stars could mean that 27 billion solar systems die per day, of a comparable size to our solar system; using Drake's equation around 1350 of these would have, at some stage in their history, supported life at least as intelligent as modern-day Homo sapiens. This using only 10% of stars, the estimated number of medium-sized stars. The Universe is a glorious and violent place.

I have viewed considerably lower scientific estimates; however, the lower figures are nevertheless considerable and support the conclusion that the death of stars is routine. How large the numbers of star deaths are, per second, is the only unknown.

The accuracy of these equations assumes a high production of stars over time; the Universe was at its brightest more than two billion years ago when scientists believe there was optimum stars in existence.

Medium-sized stars a similar size to our Sun, with a similar lifecycle, would have needed to have been formed

CHAPTER 7

around ten billion years ago, to be dying natural deaths today. These estimates therefore rely on star number production being high, around ten billion years ago.

James Vincent, the Verge Science, believes that this was a fertile period for star production in our galaxy and others. If extrapolated, it suggests star production was high enough to support incredible star deaths in the contemporary Universe. Scientists believe that galaxies of a similar size to the Milky Way formed ten times more stars 10-11 billion years ago than in modern times. The Universe was considerably more active, with greater volumes of unattached materials left over from the Big Bang in the form of gases, perfect for star creation.

The number of stars has fallen since the Universe was at its brightest, and therefore the number of medium stars with a life of an average of ten billion years, will have margins of inaccuracy. Stars supporting intelligent life are also reliant on Drake's estimate of one in 20 million and the estimate of star numbers being reasonably accurate. However, the precept remains true, that either in the past, present or future, that a number of stars divided by the average lifecycle, results in an average number dying during that period of time.

We can assume with reasonable accuracy, that a higher number of larger sized stars died each day around two billion years ago as many more were produced, than today, as they have shorter lives measured from a few million years to tens of millions of years. Whichever figures we use from high or lower star estimates of total stars, we can evidence that the death of our star will be hardly noticeable in the grand scheme of the Universe, when its time of death arrives. This death will be as significant to the Universe as the death of an insect, to life on Earth.

Contemporary humans are unable to observe this level of star destruction across the Universe. When we observe newborn stars, for example, a similar mass to our Sun, at five or ten billion light years distance. The first would not be the blue-white newborn star we are observing, but a yellow star, about halfway through its life cycle, similar to our Sun. The second would be deep red and greatly enlarged, incinerating and swallowing planets and close to imploding.

The first stars formed around 13.5 billion years ago, but the vast majority, much later. For Hubble or James Webb telescopes to observe the death of a medium-sized star, similar to our Sun, it would have to be old enough and close enough, within about 3.5 billion light years. Given the Universe is 93 billion light years across, and the vast majority of medium-sized stars observed will appear be less than ten billion years old, even though many may be dead and replaced by now. The optimum production of stars culminated two billion years ago; this greatly reduces the opportunity to observe this phenomenon.

It is worthy of note, that the galaxies we observe at 12-13 billion light years from Earth, are now probably around 90 billion light years from Earth. The effect of galaxies travelling faster than light for some time after the Big Bang.

CHAPTER 7

If we look far enough across space, we may see a newborn star, a similar size to Earth, that no longer exists as it is far enough away to have lived its natural lifecycle and imploded. The light from our Sun will still be travelling across space, billions of years after it has imploded. Indeed, the light from the stars that formed the chemical elements that made us, before our Sun existed, and now long dead, will still be travelling across space, observed billions of light years away.

This light may never reach the galaxies on the edge of the Universe, as they travel fastest away from us, at possibly close to, or in excess of speed of light. The indication that the Universe is 93 billion light years in size, in just 13.8 billion years, indicates that galaxies, at a time in the past, travelled faster than the speed of light, as mentioned earlier, but important to note.

We know the speed of separation of galaxies well within the boundaries of the Universe, started to accelerate around five billion years ago. The stretching of distance between galaxies is accelerating and is getting commensurately greater; eventually the skies will be in darkness beyond our own galaxy, by then combined with Andromeda, in many tens, or possibly hundreds of billions of years, we will be isolated in the Cosmos.

Large and medium-sized stars have a relatively rapid turnover of between a few million to tens of billion years, compared to much longer-lived small stars. The majority of stars in the Milky Way are relatively small in size, the very small ones around 10% of the mass of the Sun, or less, and undisturbed, will burn slowly for possibly trillions of years, using much less fuel. However, in a violent Universe, this doesn't mean they will all exist for hundreds of billions or trillions of years.

Scientists estimate a possible 100 million black holes in the Milky Way alone; this potentially means that billions possibly hundreds of billions of stars, of all sizes, have imploded and been consumed over billions of years, in a single galaxy. In hundreds of billions, possibly trillions of galaxies, the same story is likely to have been played out. The numbers are beyond all scale humans could have ever imagined, just 30 years ago.

Small stars that exist in binary or multi-star solar systems can be destroyed by activity or death of a larger resident star. Quasars, neutron stars plus supernovae in close enough proximity also have the power to destroy stars. The conclusion is that possibly trillions of stars, of all sizes, die every Earth day across the vastness of our Universe.

In Universal terms, the death of a solar system is routine and mundane, all life lost counts for little. When our Sun finally implodes under the weight of its own mass, it will be one of many that died that day. I started this book discussing perspective – these figures give me a new perspective, on our place in the Universe.

The vast majority of humans will not have a view, unaware of the scale of Universe they reside in. Considering the size

CHAPTER 7

of the Universe scientists believe exists beyond our galaxy, the scale of past human exaggeration regarding its own place in the Universe becomes clear. We are only relevant to ourselves. For further context and insight into how the human brain is easily influenced, consider the following.

Around eight billion people share our planet, the average life expectancy according to 'Worldometer' is 73.3 years or 26,773 days. Therefore, an average close to 300,000 humans will eventually die each day on our planet, or seven every two seconds. Seventy-three years ago, the average life expectancy for humans worldwide was 46 years, within a population of 2.5 billion. It is estimated that an average of 141,000 died per day at that time. The organisation 'World Population Review' estimate 171,000 die per day at present, around two per second. This takes account of the rising birth rate since 1950 and the instance of premature death across the globe.

This demonstrates the effect that tapered numbers of births and deaths has on averages of stars or humans. If the Earth's human population had been constant at around eight billion people since 1950, living an average of 73.3 years, then average deaths per day would be nearly 300,000. This reflects the tripling of the Earth's population since 1950, 74 years ago, meaning that the majority of people born in that time, have not yet reached the 73.3 years average. The same principle applies to average star life and death.

The vast majority of these deaths have no impact on our daily lives, yet the loss of a few human lives, of people we have never met, covered by national media, can impact our perspective, even change our views and character. Our perspective is controlled by our senses, our senses are controlled by the information fed to them. We may be concerned about a child murder, yet the other 170,000-plus deaths that day, raise no more concern than the death of an insect, in the minds of most humans.

I am not saying it is wrong, not to care about issues or people that don't affect us directly as it would be a very stressful and upsetting existence, if we did. Many do, however, and I admire them; it must be upsetting for them when others don't.

In the late 1960s as a child, I was very badly affected by famine and starvation in Biafra, since I have been horrified by senseless killing and unnecessary death. The Dunblane shootings of 1996, in particular, as the children were the same age as my son, I openly wept in front of friends. I now have a perspective changed by time and experience. I can't save the world, and worrying about it appears senseless and personally destructive.

The additional death count in World War II, worldwide, was around 75 million. Between 1st January 2024 and 15 March 2025, 75 million humans will die in less than a quarter of the six years of the costliest conflict in history in terms of human life. The increase in the world population since World War Two results in more people being likely to die in 15 months than all those who died in six years caused by World War II, 1939 to 1945.

CHAPTER 7

The estimated human population, worldwide, in 1939-1945 was approximately 2.3 billion. This rose to 2.5 billion in 1950 with 141,000 deaths per day, so then it is reasonable to assume that deaths per day between 1939-1945 was around 130,000 plus war dead. War dead of 75 million over six years, approximately 34,250 per day, making average worldwide deaths 1939 to 1945 of approximately 165,000 deaths per day. Still a lot less than human deaths per day today, averaging 171,000 and due to rise significantly in future.

In 2021 the World Health Organization recorded that eight of the top ten causes of deaths in low income economies were in categories related to premature death such as respiratory infections, malaria and birth asphyxia / birth trauma. Arguably that level of tragedy is comparable with World War Two, as many have and will die prematurely, as they do in wars. The lack of information and the lower significance of the countries suffering most, enables in the human brains in westernised countries to shut this out. A train crash killing 50 people will be news headlines, possibly for days. The death of one Royal family member or a pop star, weeks.

As with the Universe, its wonders and incredible scale, the human mind, through lack of knowledge and in many cases a lack of will to learn, places issues affecting its own planet and humanity behind a screen of ignorance, in favour of what its senses are fed.

God made the Universe and humankind is special, made in 'his' own image – this is what our minds are fed; it appears nonsense. When the Sun's fuel exhausts it will be destroyed, without ceremony, and all life in the solar system will have died, possibly hundreds of millions of years earlier. The God of Gravity will claim another victim and will continue its powerful influence for trillions of years. The Gods of humans will be no more.

ARE WE ALONE?

Earlier in this book I suggest a 'what if' there was an average of one planet with intelligent life forms, at the level of human intelligence or above, in every 20 billion planets, that would suggest at least one trillion planets of that type within the Universe. A loaded equation to arrive at a trillion planets sustaining intelligent life. This figure was predicated on one intelligent life supporting planet for every 20 billion stars.

If we adopt a more scientific approach, we find that a trillion is actually a very conservative estimate. In 1961, as referred to earlier, Frank Drake created a formula called the Drake Equation. This exhibits a more scientific methodology for estimating the presence of intelligent life beyond our solar system.

The equation is complex and heavy for this book, but simplified it includes estimates of star formation rates; the ratio of stars to planets; the average number of planets that can eventually support life per star that has

CHAPTER 7

planets; the fraction of planets that could support life at some point; the fraction of planets with life that go on to develop intelligent life and civilisations; the fraction of civilisations that develop technology that is capable of releasing detectable signals into space.

Drake arrives at an estimate, which combined with contemporary information suggests 10,000 stars supporting intelligent life on at least one planet across the Milky Way. Using an average estimate of stars in our galaxy of around 200 billion stars (estimates vary from 100 billion to 400 billion), that equates to around one star in 20 million has a planet with intelligent life.

Drake was unaware of the vastness of the Universe the Hubble Space telescope later discovered. By applying his equation on a low estimate of 2×10^{22} stars, extrapolating his equation across the Universe, results in an estimate there could be about 1,000 trillion planets that potentially host intelligent life. If the actual number of stars exceeds this 2×10^{22} figure, which is probable, the potential number of planets that could support intelligent life may increase by thousands of trillions, when using higher estimates.

The higher estimate of a septillion stars would result in an extrapolation figure of 50,000 trillion (5×10^{16} or 50 quadrillion) stars with planets sustaining or having the potential to sustain intelligent lifeforms. If the correct number is only a small fraction of this number, it still means that trillions of intelligent life forms, at least as intelligent as humans, exist.

I imagine communicating with these extra-terrestrial life forms and trying to explain that if there is a God, it is Earth and humans that are the chosen ones, we are in 'his' image. Then I realise, how ridiculous and condescending that would sound, probably because it is.

Of note is that if the Drake equation and estimate of stars by NASA are accurate, and making the rash assumption that 10% of stars are an approximately similar size to our Sun, then in excess of 13,500 worlds are on the same Earth day of age as our Earth relative to the age of their star; different sized planets spinning at different speeds will obviously have different day lengths. This assumes that these planets were created near the start of the solar system, and we can reasonably measure time in Earth days, estimated by dividing 50,000 trillion by the number of days in ten billion years.

Many habitable planets will be developing varieties of early life now, as Earth did, and many coming to the end of life, as their star expands towards death, as our Sun will, inevitably consuming Earth and any remaining life. Not included in the estimated 50,000 trillion habitable planets, those long dead planets, with intelligent life now extinguished, and tens of thousands of trillions, that will exist in future. We are in a tiny moment of time in our Universe's history and future. To believe we are in some way unique, defies all logic.

From a young age, I was taught that God is the creator of the Universe, and 'he' fashioned humanity in 'his'

CHAPTER 7

image, making humans the 'chosen' species. This 'Drake based' estimate of intelligent life appears at odds with that assessment. Now that we have this scientifically produced estimate of intelligent life, will we continue to believe that humans are in some way 'special', created in God's image and regarded as 'his' chosen ones? With emerging knowledge, I find this assumption deeply flawed, self-indulgent and a nonsensical conclusion. Yet an easy mistake to make if you were having a guess 1000-2000 years ago.

If Drake's calculations are highly inaccurate and my much lower guess of one in 20 billion stars supporting intelligent life on a planet is used, there is still plenty of scope for his equation to be a vast overestimate yet still provide for the existence of trillions of planets with intelligent life.

NARROWING THE SEARCH

Drake's proxy estimate of 10,000 planets with intelligent life in our galaxy may sound like a promising target to aim at to contact another intelligent species. One in 20 million is the sobering reality. Using Drake's estimate, investigating the first ten million stars and their multiple planets would give a 50:50 chance of discovering intelligent life, plus the need for contact to be reciprocated, to confirm. If the intelligent lifeform is 40 light years away, then it would take 80 years to send and receive a message. A human lifetime.

Earth is 75,000 light years from the far edge of the Milky Way, 150,000 years for a send and reply message. The Milky Way is more than 100,000 light years long, 1000 light years travelled would be less than one per cent, and only in one direction, therefore a miniscule part of the galaxy.

The vast majority, possibly all, of intelligent life in our galaxy is likely to be out of reach across a single generation; the hunt is a multigenerational task, yet it has to be started by one.

What is the likelihood of life existing within a relatively close distance to Earth? Twenty out of the nearest 30 stars to Earth are thought to be Red Dwarfs, including our closest star; three quarters of the stars in our galaxy fall into this category, some estimates as high as 85%. These stars failed to form properly with insufficient energy to become stars like our Sun.

The smallest red dwarfs being under ten per cent of the size of our Sun, the coolest having a surface temperature a third of our sun at a mere 2000 Kelvin, about 1700 Celsius. Our Sun has a surface temperature of 5600 Celsius, close to 6000 Kelvin. The variety of stars, planets and their relative imperfections is immense, a very high percentage unable to support life. Yet they exist for possibly thousands of times the life of our Sun, which under the right circumstances would increase the chances of supporting life, at a time, within that incredible length of time.

CHAPTER 7

Red dwarfs are the 'slow burners' of the Milky Way. This goes some way to explaining why they are the most common. Many will have existed almost across the history of our galaxy with possibly hundreds of billions of larger stars having been created and died in that time. Small red dwarf stars can exist for hundreds of billions, possibly trillions of years – given their abundance and proximity, it is important to know if they could sustain life on orbiting planets. They would, however, have to avoid the dangers of a highly volatile universe to achieve that length of life.

When you gaze at the night sky, a far greater number of stars remain hidden from view than those that are visible within the observed area of space. Many of these 'difficult to see stars', are red dwarfs. Of those red dwarfs observed by telescopes, one in 20 has at least one planet the correct approximate distance from the star to support life, referred to as the 'Goldilocks zone'. Because red dwarfs have much lower surface temperatures than larger stars, the Goldilocks zone, suitable for supporting life, is much closer to the star. Some research estimates 40% of red dwarfs could have exoplanets in habitable zones.

On the one hand, the closer necessary proximity would give increased hope of life on the nearby planet, the cooler the star, the closer a planet has to be to support multicellular life. However, proximity causes other problems.

This necessary closeness to a red dwarf star in order to support multicellular life is highly likely to cause the planet to be in a 'tidal lock' to its star. Our Moon is tidal locked to Earth's orbit, because of close proximity, which means its rotation is slowed by gravitational forces and only one side of the Moon is observable from Earth, as it is held by the Earth's gravity.

If not in a tidal lock, a planet, the necessary distance from a red dwarf star, would have potentially trillions of years to develop life. Red dwarfs have access to all their hydrogen for producing energy, large and medium size stars don't, and burn fuel at a very slow rate, compared to other stars – 2000 degrees Kelvin would be enough to support life in close enough proximity.

If our Sun's surface temperature was at this level, Mercury and Venus would be cooler; however, the Sun's gravitational effect on these planets' revolutions, means that a day on Mercury is over 1400 hours and on Venus over 5800 hours. Both mitigate against an environment that would host life above unicellular form. Although not tidally locked, the same forces are slowing the rotation of these planets.

A planet that orbits a red dwarf star, maintained in position by its star's gravity, tidal-lock, will feature one side heated intensely by the star and a contrasting side that remains extremely cold, facing away into space. Imagine holding a 'selfie stick', with a mobile phone attached, and moving it around your head with the screen facing you at all times during a 360° rotation. The stick holds the camera in place in a similar way to

CHAPTER 7

how gravity holds the moon in place or a tidal-locked planet is held in place by its red dwarf star.

Multicellular life as we know it probably cannot survive on either side of a tidally locked planet. However, the extended life of a red dwarf means that this could possibly change with a high energy event, such as the planet getting moved further enough away to break the tidal lock but still be able to support life.

Our Moon can have temperatures of 120C on the side facing the Sun at any given time and -240C on the dark side, in the bottom of deep craters.

The possibility for life on a tidal locked planet lies on the border between light and dark, a twilight area which is permanently dusk; this is the best chance of having temperatures conducive to supporting life. It could potentially have three types of H_2O in steam, ice and water, as our planet does. All three contain hydrogen and oxygen; in addition, carbon is abundant throughout the Universe, and together these form the basis of life as we know it. Single cell life would seem a distinct possibility, complex life less likely unless tidal lock is broken and planets can spin at the necessary speeds on their axis.

In conclusion, it is unlikely life will be discovered on planets relatively close to Earth, unicellular life, even life to the level of humans 100 years ago on our planet, would have been difficult to detect from distant space, as we didn't have capacity for interstellar communications.

The Drake equation combined with the types of stars in Earth's vicinity makes it statistically, highly unlikely.

The Kepler space telescope could observe 150,000 stars simultaneously for ten years until it was decommissioned in 2018. Any variation in brightness of a star would signal an object orbiting, size and distance from the star could be calculated, and exoplanets, planets outside our solar system, could be confirmed.

Kepler discovered more exoplanets in its almost ten years' operational life than all the telescopes on Earth during that same time. Planets likely to host complex, multicellular life seem to be uncommon, which is hardly surprising considering the Universe's inhospitable nature.

The Earth itself was inhospitable to multicellular life whilst hosting single cell life for more than two billion years. Planets like Mars and some of the moons orbiting gas giants like Saturn and Neptune may contain water under their surface and possibly life.

Evidence of water once existing on the surface of Mars are clearly visible in dried up rivers and oceans. An ocean once over a mile deep is believed to have been present in the northern hemisphere of Mars. Life almost certainly had suitable conditions on the surface of Mars at a time in the distant past.

Planets with a possibility of multicellular or even highly complex life similar to that on Earth, are being selected

CHAPTER 7

for extra scrutiny by scientists using instruments such as Hubble and James Webb telescopes.

Our own galaxy is of course the starting point in our search for complex life. Its proximity and numbers of 100 to 400 billion stars, is enough to keep scientists occupied for the foreseeable future, but improving technology will enable this search to be expanded and accelerated. Efforts to place a more accurate figure of stars in the Milky Way is hampered by incredible numbers and also that most of the stars will be Red or Brown Dwarfs which do not burn brightly, making them more difficult to locate.

This exploration has now confirmed that billions of exoplanets are likely to exist in our galaxy and therefore billions of small chances of life, if added together, raise the possibility of life considerably.

We then need to consider two more factors. Firstly, hundreds of billions, possibly trillions of galaxies exist, existed or will exist, in the Universe, meaning life is likely to, has, or will exist on trillions times trillions of exoplanets, and most probably, complex and intelligent life on many thousands of trillions of planets across the Universe. Incredible distances being the barrier to us ever confirming this estimate.

Secondly, many smaller stars will outlive our Sun, providing a greater potential for life; they can emit life-supporting heat for hundreds of billions or even trillions of years, which greatly enhances the opportunity for life to arise.

In summary, large stars die young, so these are therefore unlikely to produce planets with life, which on Earth took more than 500 million years to produce in its most non-sustainable, basic form. It took over 800 million years from the formation of our medium sized star, the Sun, to achieve sustainable single cell life on Earth and three billion years to form multicellular life; many stars don't live that long. A star of two solar masses is likely to have a life cycle of around 2.5 to 3 billon years, much less than our Sun.

Smaller and cooler stars are less likely to support life because of the necessary proximity of planet to star and the affected gravitational relationship. The 'slow' burning stars are likely, however, to live for hundreds of billions of years, increasing opportunity for conditions for life to be met.

As discussed earlier, scientists believe the Universe hit its maximum brightness around two billion years ago; before this point more stars were being formed than dying. This is no longer occurring and the number of stars formed in the Universe in the last 13.8 billion years is likely to be more than those that ever will form.

That number yet to form will still, however, be many more than exist today. A galaxy the size of our Milky Way will produce an average of seven new stars per year;

CHAPTER 7

11 billion years ago it would have been ten times that. Scientists believe the Universe has lost around 50% of its brightness in the last two billion years.

The shorter life cycle of larger stars will have proven essential to the creation of life around smaller and medium sized stars, as life-giving elements are formed within the dying star. Large stars live and die quickly, making the formation of life-giving planets around them unlikely. They have, however, supplied the essential materials that sustain life in other parts of their galaxy, with an abundance of elements found in space debris and gases. This enables a supply of necessary elements for life, that can be blasted through space after the death of these larger stars.

Our Sun will have a total life of about ten billion years, but may only provide the necessary environment for life on Earth for between 60-80% of that time, as little as 30% for complex or multicellular life. The two billion years head start that single cell creatures had on multicellular creatures contributes to this difference, as will the longer continued endurance of unicellular creatures, long after multicellular life is extinct, sometime before the Sun eventually expands and, finally, incinerates Earth.

Single cell creatures will survive until much closer to the death of our star than more complex life, some being able to tolerate temperatures of well over 80C, up to 95C. Scientists accept that multicellular life on Mars is highly unlikely, because of the harsh conditions, yet consider the possibility of single cell life under the surface of Mars. Earth will eventually look similar to Mars with its life-protecting atmosphere removed, but single cell life could possibly endure below Earth's surface for many millions of years.

In summary, an ageing star brings with it extinction of life, where life exists, complex and multicellular being most at risk, earlier. Single cell creatures like bacteria lived for possibly billions of years before multicellular life, in environments that could not possibly support multicellular life, similar to Earth, when the Sun expands, and will still be present on Earth possibly billions of years after multicellular life is extinct.

The number of stars known to exist, make intelligent extra-terrestrial life an inevitability. Planets resembling Earth will likely exist, and in great numbers. Others will have existed and died, but for every Earth-like planet in the Universe there will be millions or possibly billions that didn't take the same route of development because of their size, constituency, proximity to a star, plus many other reasons.

In the past 13.8 billion years, possibly many septillions of stars and planets have already lived and died in that time, an almost unimaginable level of variety across the Universe. More time is yet to come, the happenings of the future, which is highly likely to be much longer than the past.

The Universe houses immense forces that constantly drive change, shape its structure, create and consume

CHAPTER 7

stars, release energy across vast distances, generate life, and determine its end, while maintaining the fabric of existence. Provision of 'God-like' powers in creating and destroying on a mass scale, with a by-product of life. What are the objects and forces which have God-like powers as described by humankind for millennia.

*

References, sources and recommended reading or research, Chapter 7

How the Universe Works – Pioneer Productions / Discovery Channel, 1.1, 1.2,1.8,4.5 6.5,6.7,7.4,7.8,8.4 & 11.3

Matter and energy are interchangeable – Einstein's Big Idea, pbs.org>Einstein

The new mystery hidden inside the universe's biggest ever black hole - BBC Science Focus, Dr Becky Smethurst

How many atoms are there in the human body – education.jlab.org>mathematics

The star lifecycle – webbtelescope.org

Are planets tidally locked to red dwarfs habitable? It's complicated – Evan Gough, Phys.org

Chemical evolution, astronomy – swin.edu.au

How stars die and how long do they live – Sky and Telescope, Maria Temming

The Milky Way, imagine the Universe – NASA, gsfc.nasa.gov

Types of black holes – NASA Science

CHAPTER 7

Red Dwarfs: The most common and longest lived stars
– Space.com

Kepler /K2, NASA – observations of stars with potential life supporting planets

Andromeda – Milky Way collision will see our galaxy merge with its neighbour – BBC Sky at Night Magazine

Awakening newborn stars – NASA Hubble Mission

The Stars within us – Jason Bates, National Science Foundation

Micro Universe inside an atom – Stable Diffusion online

Sagittarius A Star: The Milky Way's supermassive black hole

The life-cycles of Sun-like and massive stars – Webb Telescope

Casualties of World War II – Lumen Learning

Second World War Deaths – Statistica

Star Basics – NASA

Galaxies are moving away at an increasing rate, carried by the expansion of the Universe – Alastair Gunn, BBC Sky at Night

DOGS THAT EAT GODS

***Stellar Nucleosynthesis: How stars make all elements, how elements from hydrogen and helium are created** -*
Andrew Zimmerman Jones, ThoughtCo.

CHAPTER 8

The Universe's known Candidates for Gods

—

On first presenting this book to my publisher, I was described as an atheist. Yet I do believe in gods of creation. I suppose, I can conclude, that the difference between myself and a 'traditional' believer, is that I don't believe in invisible, intelligent gods. My gods are visible, or at least recordable, and highly influential, proven to exist and are undoubtedly responsible for everything living, plus every inanimate thing we can see, touch, hear, smell or taste.

The known Gods of the Universe are interventionist but on an incredible scale. They are instrumental in creation and destruction on a scale beyond human imagination, holding matter together, controlling spacetime and creating galaxies and stars that are the very fibre of the Universe.

They don't bother themselves with which side of insignificant species will win a war on a tiny piece of rock in a small to medium sized solar system, in an unimpressive common type of galaxy, controlled by a relatively small black hole. In around five billion years our solar system will be destroyed, human gods long gone with it, and the supreme powers will continue towards infinity.

Atheism is a view from your own personal hilltop. Romans considered Christians atheists because they didn't believe in Roman Gods. Those who practise modern day monotheism, Christians and Muslims, for example, do not consider those who practise polytheism as atheists, or vice versa, presumably because they believe in intelligent, interventionist gods which get involved in controlling minor relatively insignificant details in the context of our Universe.

'There are four thousand Gods, you don't believe in three thousand, nine hundred and ninety-nine of them, I don't believe in one more' Ricky Gervais, Afterlife.

The Universe is likely to exist for trillions of years, long after our Solar system and even our Galaxy no longer exist. Humanity and the many concepts of God it holds will have been a blink of an eye, in comparison to time elapsed and that going to elapse in future. Given the numbers of intelligent lifeforms that are bound to exist throughout the Universe in the present and into the future, gods of some description are likely to exist throughout the Universe and for a very long time. Whether they are invisible, omnipotent Gods, in the

CHAPTER 8

majority of cases, is doubtful. As intelligent life becomes more intelligent, will today's contemporary gods still exist? I believe not.

Gods of Rome and of the Incas and many, many more are testimony to temporary gods, which all gods appear to be. Many contemporary religious devotees will deny this, but the evidence supports the natural end of religions, time is the only unknown. To have told Romans, Greeks or Norse that their Gods are temporary would have extracted the same levels of protestations and indignity that informing a Christian or follower of Islam would today.

The immense scale of the Universe presents abundant opportunities for complex life to evolve. Intelligent beings on different planets will likely develop their own unique understandings of gods and supreme entities, these will all end at some future time and long before their star consumes their planet. The lucky ones may be able to travel to a reachable planet and begin again, only to delay the inevitable.

The idea of gods, as understood by modern humans, dates back approximately 6000 years, possibly longer. The following of religion may have reached its peak and human belief in God appearing to be in decline, in Western developed economies and forming cultures. As discussed elsewhere in this book, humans gained the capacity to believe in gods sometime in the last 10,000 years, they now have the capacity, environment, opportunity and culture, not to believe, if they so choose, as human intelligence and the questioning brain accelerates at an unprecedented level.

Better and regular nutrition developing brains, information highways, discussion, freedom of expression are all fuelling this change. The compulsory adherence to religious teachings and practice has come to an end across enlightened, free, democratic societies. It is very possible that humankind's exposure to God could last as little as 8000-12,000 years in total, if the belief in God dies out within 2000 years.

The Information humans have now, no writers of any religious texts in history were party to. Consequently, they lacked the understanding needed to follow modern reasoning. They may not have concluded the same as I have, but I'm sure their conclusions would have been very different to those written in the past. Through brainwashed generations, who had little choice other than to follow and believe, as the alternative to belief in religion would result in being ostracised, banished or killed. Religion of just a few hundred years ago seems very similar to a mind-controlling dictatorship.

Without alternative sources of information and knowledge which contradict information already provided as knowledge, how could people believe anything other than what they were taught from birth? An entirely believable narrative based on the available information at that time.

CHAPTER 8

With emerging information evidence pointing us to alternative conclusions, God, as described by ancient humans, is being doubted. If there is no 'human conceptualised Gods', there is no shame in being wrong – humans didn't know much of the information we now know about the Universe, until well into the lives of many humans currently living. In the past 30 years, through the Hubble space telescope and other instruments, the age of the Universe, the scale has been discovered, and with it the insignificance of humankind, against this scale of time and the size of the Universe.

In the past, one set of adjoining generations had to cease believing in now 'extinct' gods such as Roman, Greek, Norse, Egyptian et al, resulting in those religions, and Gods, no longer existing in the minds of humans. In this information age of free debate and expression, I believe we have entered the first of a set of generations that will banish 'invisible' gods in favour of forces that are known to exist and have the power many religions believe Gods possess. People will not worship these, but rather marvel at them.

We know that the forces of the Universe have been driving creation and destruction for 13.8 billion years, more than 13.7 billion years before humans conceptualised God. Forces that possess energy and power of trillions of times that of our Sun. The power to slow and even stop time, to bend spacetime and to release more energy in a second than our Sun will release in its 10 billion years' lifetime. The power to hold everything in the Universe together and stop all matter becoming sub-atomic dust. You are about to be introduced to the 'known gods' – give them the respect they have earned.

CANDIDATES FOR THE GODS OF THE UNIVERSE

Black Holes

'Black holes help build galaxies', *How the Universe Works*

Through the Hubble telescope findings and other observations, scientists now know our Universe to be a purveyor and overseer of mass destruction and creation. If it wasn't, Earth and its life would not exist, as discussed earlier, the chemicals and metals that make our bodies were created in dying stars probably billions of years before our solar system existed. We can conclude that this destruction is the basis for renewal and recycling of the building materials of stars and planets.

The reuse of this material may take billions of years to come about, but that is a mere blink of an eye, to a Universe likely to exist for trillions of years, around 100 billion years of that time, continuing the creation of stars, planets and in some cases, life itself. Professor Brian Cox estimates that the Universe is unlikely to support life for more than 4% of its existence.

CHAPTER 8

The most efficient destruction units, and therefore creation renewal units, in our Universe are ultra-massive black holes followed by supermassive black holes. These are usually situated at the centre of galaxies and have a gravitational influence on the Galaxy. Without them scientists calculate that Solar Systems within a galaxy would drift away into isolated and distant space.

Isaac Newton, Albert Einstein and others taught us that the strength of gravity is related to the size of mass. The larger the mass, the stronger the gravity. Sagittarius A star at the centre of the Milky Way, is the largest single mass in the Milky Way, and exerts an influence on our solar system which over time will pull us closer, but no panic yet as it is 26,000 light years away from our solar system. Our supermassive black hole, although all-powerful in our Galaxy, is relatively small compared with many other black holes at the centre of many other galaxies.

Fifty-four million light years away lies M87, an ultra-massive Black Hole, 6-6.5 billion times the mass of our Sun. We are observing this mass 54 million years ago and it is likely larger by now. Over the past two to three years, black holes weighing in at 20-30 billion solar masses have been discovered.

The largest known object in the observable Universe being in Ton 618 Galaxy, with a black hole of estimated to be at 66 billion solar masses, or up to 22,000,000,000,000,000, 22 quadrillion (22 million, billion times) the mass of planet Earth. It is likely to be considerably wider than our solar system in size, from the Sun, beyond the furthest planet. Some estimates suggest that the radius measures twenty times the distance from the Sun to Pluto.

The frequency of recent discoveries illustrates the rapid pace of exploration in our 'knowledge age', thanks to space telescopes that penetrate obstructive clouds of space dust using infrared and other technologies. This allows scientists to observe the Universe with greater detail than the Hubble space telescope can provide. It was as recently as May 2022 that the first images of a black hole were first confirmed; it was 1964 when the first evidence of black holes existed, yet images had remained elusive until recently. In 1915 Einstein theorised about the existence of the incredible phenomenon we now describe as black holes, showcasing his genius to the modern world, now they have been confirmed.

Science continues to demonstrate its brilliance, with black holes theorised and now discovered, together with Einstein's theory of gravitational waves. Science deals in facts, theories, and balance of probability given available evidence, not superstition. Recent human history has for the first time given us a window to creation and how the Universe works. Scientific theories should not be underestimated as many are proven correct over time. The opposite could be said of much theology and religious doctrines.

A black hole forms when a star (normally over 15 solar masses, 15 times the mass of our Sun), collapses, usually at life's end, when it has burned its fuel supply. It collapses

CHAPTER 8

to a size many thousand times smaller than its original size, but maintains much of its heavier mass as gravity crushes it with massive force.

The size of our Sun prevents it from becoming a black hole; instead, its collapsing mass will likely generate a smaller white dwarf star that will ultimately extinguish.

Black holes are usually measured in three sizes, black holes up to 100,000 solar masses, supermassive black holes, 100,000 to five billion solar masses, and ultra-massive black holes, over five billion solar masses.

The Universe probably contains trillions times trillions of black holes spread throughout its hundreds of billions of galaxies. Scientists recently discovered that black holes likely merge with one another or with supermassive black holes to create even larger black holes, thereby enhancing their gravitational pull relative to their increased mass. They truly stand as the dominant heavyweights of the Universe and its galaxies.

Black holes have such a strong gravitational pull that they can stretch and compress atoms, breaking matter down into sub-atomic particles. An atom is composed of 99.9999999999996% empty space, with its mass primarily derived from the nucleus's protons and the orbiting electrons.

Crushed and pulled apart under extreme pressure, this means it may be possible to crush hundreds of trillion atoms into the space of one atom. This assumes all the empty space is crushed out of atoms, but electrons and protons are not crushed smaller which, given the forces at work, could happen, reducing the space needed, yet further.

Outside a black hole we know that electromagnetism is a stronger force than Gravity. Electromagnetism, which encompasses electricity, magnetism, and light, binds atoms together, giving them a solid appearance and feel despite being mostly empty space. The gravity in a black hole is many, many times greater than even the largest star can produce. Matter gets stretched and compressed; even light cannot escape the intense gravity of a black hole, which is why black holes appear black.

The capacity of a black hole with approximately one billion solar masses is beyond comprehension. If a large black hole were to swallow the Earth, extreme gravity would likely compress all the particles of our planet into a space the size of a golf ball.

When an object gets close to a black hole, the part of it nearest to the black hole experiences a stronger gravitational pull, causing the object to stretch. This draws the object into a line as narrow as an atom and subsequently extends it across the event horizon, pulling it into the black hole. Although scientists don't know for certain, it is likely split into sub-atomic particles which are crushed smaller.

The creating of the atomic bomb demonstrates the huge amount of energy stored in a single atom, the energy

CHAPTER 8

stored inside a black hole which may have consumed trillions of stars, planets and solar system sized gas clouds, is therefore beyond imagination.

We know gravity increases with mass; to put this into perspective, a human who weighs 70kg on Earth would weigh 10kg on the Moon and nearly 170kg on Jupiter and if not incinerated or crushed, 2000kg on the Sun.

This greater gravity is reflected in escape velocity – to get free of the Moon's gravity a space probe must be travelling in excess of 8,640km per hour, to get free of Earth's gravity 39,600km per hour, over 11km per second, to escape the Sun's greater gravity 2,224,800 km per hour, 619km per second. Escaping the gravity of a black hole requires a velocity of greater than the speed of light.

A smaller black hole with about 15-25 times the mass of our Sun generates enough gravity to prevent light from escaping; now consider the immense gravitational forces surrounding a black hole with billions of solar masses.

Sagittarius A*, at 4.3 to 4.6m solar masses, possesses immense gravitational force that it can consume stars, including many that are much larger than our Sun. Black holes have an insatiable appetite for stars, planets and especially gases. Over time, this immense gravitational force will attract increasing amounts of matter, including stars, planets, and gas, tearing them apart.

The black hole doesn't consume all matter it draws into its orbit, it has a very hot accretion disk surrounding it, a circle of matter around its equator travelling close to the speed of light and kept at millions of degrees Celsius.

Black holes have strong electromagnetic force radiating from their poles and surrounding the black hole. This phenomenon can draw in extremely hot and rapidly moving matter towards both its poles while propelling this energy hundreds of thousands, possibly millions of light-years, almost at the speed of light, in a bolt of intense energy, across galaxies and even beyond. A black hole has transformed into a quasar when this occurs.

A quasar is a temporary life stage of a black hole, scientists estimate that Sagittarius A star was in a quasar state around 6-9 million years ago; had its poles (beams) been facing our solar system it could easily have incinerated it, including all life on Earth. Humans and their gods would never have existed.

Quasars

Powered by black holes, some quasars are so powerful they can release more energy in a single second than our sun will release in its 10 billion years' lifetime. It was calculated that a recently discovered 'scar', an empty area of space, a million light years across, which once contained stars had been destroyed by quasar energy equivalent to one quasar producing the power of 20 billion, billion megatons of TNT, every one thousandth of a second for 240 million years.

CHAPTER 8

The scale and power of black holes powering quasars is truly unimaginable. This highly intense energy is shot from both poles of the black holes, giving it quasar status, one of the most powerful events in the Universe. Planets or stars located in the trajectory of this energy burst would vaporize, even if they are light years distant. The brightest quasar has been estimated to be 500 trillion times the luminosity of our Sun.

Energy and sub-atomic matter ejected from quasars will probably become reusable in the future for forming stars and planets. Scientists generally agree that Quasars play a crucial role in the formation of stable galaxies, contributing to their longevity.

Black holes rip apart stars, planets, and gases, and in the process of forming a quasar, they launch massive amounts of matter and energy throughout and beyond their galaxies. Materials that can be reused to create new stars and planets. A galaxy without the materials to produce stars will eventually die, therefore quasars are key to the regeneration of galaxies through providing materials for star creation. Breathing life into otherwise dying galaxies.

In achieving its immense mass and size, it is likely that Ton 618 ultra-massive black hole has consumed more matter than the total matter in our Milky Way, which may contain up to 400 billion stars, 100 million black holes and a supermassive black hole.

Black holes are certainly a God-like candidate.

Electromagnetism

Approximately a billion years after the Big Bang electromagnetism injected 'life' into the Universe by helping to create stars; without it, this process would be impossible. Gas, gravity, and compression combined to form stars once conditions became dense and hot enough to initiate nuclear fusion.

Before the star can form fully, it needs to draw in enough gas under immense pressure to cause temperatures of 15 million Celsius plus. Electromagnetism holds all matter together to enable this process. Atoms rely on electromagnetism to power electrons which enable atoms, and therefore matter, to exist and function.

When we look at ourselves and our surroundings, we may look solid enough, but as science has discovered, the atoms that make up our Universe and our bodies are 99.9999999999996% empty space. What stops us falling through the floor and tumbling to the centre of the Earth? The answer is one of the Universe's most powerful and enduring forces, electromagnetism.

A most amazing force within our Universe. If the force of electromagnetism stopped working, everything in the Universe would turn to the finest sub-atomic dust as atoms fall apart. If it were possible to travel through what was once our Universe, after electromagnetism had ended, everything that once was, would be undetectable and as mass no longer existed then gravity and spacetime

CHAPTER 8

would no longer exist as we know them. The Universe would appear as empty as it was before the Big Bang, yet it would contain the mass of the Universe it once was, but in the finest invisible dust.

Seven thousand trillion, trillion (7,000,000,000,000,000,000,000,000) atoms make up the average 70kg human body and every one of them reliant on electromagnetism to hold them together. Approximately two thirds hydrogen, one quarter oxygen and a tenth carbon, these three elements account for 99% of human atoms.

Bacteria are made up of 7% nitrogen, 14% oxygen, 29% carbon and 50% hydrogen. Humans and bacteria, a sack of similar chemicals in different sizes put together in a different puzzle. All held together by electromagnetism.

Inside atoms, negatively charged electrons orbit positively charged protons and in some cases neutrons, travelling at incredible speeds. Scientists have calculated this speed approximately 2200km per second. Put into context, that is a speed that would circle the Earth in just over 18 seconds.

Travelling at that speed and around something as small as an atom nucleus results in the electron orbiting at 7,000,000,000,000,000 (7000 trillion) times per second. About 3,200,000 revolutions would cover a millimetre. To give context to the size of an atom, a single bacterium is made up of hundreds of trillions of atoms, yet bacteria are only clearly visible to the human eye, through a x400 microscope. Incredibly small.

If you imagine an aeroplane orbiting the Earth at seven thousand trillion times per second in random orbits then the Earth would be fully obscured by the blur created. It would appear, from space, to have an outer coating of solid, unrecognisable, aeroplane. The electron gives this solidity to the empty space that makes up an atom.

There is only one force in the Universe known to be more powerful than electromagnetism and that is strong nuclear power. Nuclear power is, though, localised and relatively short lived within stars, whereas electromagnetism stretches across the Universe in every direction, possibly over 90 billion light years. It is everywhere that matter exists.

Describing electromagnetism as God fulfils many criteria outlined by contemporary humans. It is all around us, in us, it holds our world together, next to all powerful, it is omnipotent. Non-intelligent and without realisation, unlike a religious god it has no human characteristics, but is a main reason we exist.

None of the other candidates could exist, supernovae, stars, quasars, planets, moons are all dependent on matter being held together by electromagnetism; without matter, gravity would not exist or be so weak, hardly noticeable. Hail electromagnetism a true (known) titan of our Universe.

CHAPTER 8

Strong Nuclear Force and Nuclear Fusion

Strong nuclear force holds protons and neutrons together in the atomic nucleus. It is the powerful force known to science; similar to electromagnetism it also plays a critical role in holding atoms together, within its sphere of influence – without it, stars wouldn't exist. It may have its title removed in future, when more is known about dark energy and dark matter.

Dark energy appears to have the power to drive galaxies apart at increasing speed, it accounts for between 70% of all the energy in the Universe, whilst the latter is believed to account for 85% of all matter in the Universe and holds galaxies together. It likely had a role in their formation, exerting gravity in keeping galaxies together. Two potentially awesome powers of our Universe. Dark energy appears to function at its strongest between galaxies in driving them apart, whereas dark matter appears to function strongly within galaxies and exerting influence on galaxies in close proximity, bringing them together. Both remain mainly theory-based phenomena. But remember Black Holes and Gravitational waves were once scientific theory, now scientific fact.

About 100 times a stronger force than electromagnetism and is 100 trillion, trillion, trillion times stronger than gravity. Similar to electromagnetism, it holds matter together. It powers Stars through nuclear fusion and repels the gravity that seeks to crush a star. Nuclear fusion in the Sun produces the heat that supports our food growth, warms our atmosphere, and prevents our planet from reaching uninhabitable temperatures.

As powerful as strong nuclear force is, it is a localised power, and relatively short lived in stars. Electromagnetism and gravity are in place throughout the Universe and have almost eternal properties that will exist long after the last star has burned itself out. However, with its immense power and life-sustaining properties, it remains a worthy nominee for a creator, life-giver and life taker.

Time

Einstein showed us links between time, space and gravity, but what is time? Time is 'progression' relative to gravity and therefore mass. Einstein explained that time moves more slowly with proximity to greater mass, greater mass exerts greater gravity, which slows time. One of the most challenging concepts for an uninformed person is to grasp that time passes at varying rates depending on specific locations in space. From birth, the vast majority of humans view time as a constant, yet it isn't.

I prefer to view time as a continuous movement into the future, regardless of how fast the clock hands travel. Many discussions and theories have existed regarding time travel – it may be possible to move forward in time, but scientists are generally in agreement that backwards movement in time is not possible. We cannot logically go back to a dimension where time has already passed.

CHAPTER 8

It must be assumed, therefore, that travelling into the future would not be reversible.

Yet logically, moving to a larger mass would effectively be moving back in time, as less time has elapsed, whereas moving to a smaller mass would be moving forward in time. Did astronauts move into the future with the moon landing as time moves faster on the moon and then move back in time on return to Earth? Truly a mindboggling consideration. Hardly noticeable because of the extent of differences in mass. Theoretically, a huge difference if travelling to a black hole and back to Earth. My inferior and sometimes confused human brain can only conclude that we can't travel back in time within the world we occupy, at any instance in time.

Humans have defined time on Earth based on motion. The Earth rotates around its axis, and we refer to one complete revolution as a day. The Earth completes one orbit around the sun in approximately 365 ¼ days, which we define as a year.

Various planets throughout our solar system, galaxy, and the Universe rotate at different speeds. A complete rotation on Neptune takes approximately 16 hours and six minutes in Earth time. Neptune is much larger than Earth, but it rotates on its axis at a much higher speed. Neptune completes one orbit around the Sun in 165 Earth years. No human has ever celebrated their first birthday in Neptune years. A huge potential saving on candles over a human lifetime.

As the scientific consensus is that we can't travel back in time, time is a forward not a backwards progression, regardless of its rate and which galaxy, star or planet it ticks over. It is, however, a progression at different rates, not constant as it appears on Earth. Rate of time is related to mass, the closer to large mass the slower time progresses.

When humans hear the phrase 'millions of years', they typically conceive of it as an almost unfathomable timeframe. For the Universe, a million years is merely a brief instant in its journey that is anticipated to last for trillions of years. Although scientists believe humans exist 13.8 billion years after the start of the Universe, we are here almost at its relative start, compared to what time is likely to elapse, in future.

The huge mass of black holes will bring time to a standstill on its event horizon, gravity unleashed, yet progression continues, the black hole continues to absorb matter faster than the speed of light.

Time moves more slowly at the bottom of the Mariana Trench, the deepest point on Earth's surface, in the Pacific Ocean, compared to the summit of Mount Everest. This occurs due to the proximity of the Mariana Trench to Earth's centre of mass. This amounts to minuscule fractions, barely perceptible throughout a human's lifetime. You would notice this difference more clearly close to the Sun and even more dramatically nearby a black hole, where time might appear to stop.

CHAPTER 8

Activity unfolds rapidly near a black hole as it consumes stars, planets, and space dust within an accretion disc that approaches light speed. It appears that progress flows steadily, despite the stillness of time.

The Earth has gradually decreased its rotation speed on its axis over billions of years. Experts debate the extent of the slowing, but scientists agree that it is happening. Some estimates suggest it slows by roughly 40 minutes each day over the course of 100 million years. Three hundred million years ago, an Earth day lasted 22 hours, and in another 300 million years, it will extend to 26 hours. Although some other estimates are that a day was 23 hours in duration as long ago as 1.2 billion years ago. Science agrees that there was a significant difference, but timescales and rates of slowing are still debated.

If Earth's orbit around the Sun slows down, a year will become longer. If it moves closer to the Sun while maintaining the same speed, then the years would become shorter. Human measures of time merely express a unit of progression which is fluid throughout the Universe, not constant.

The Universe may be 13.8 billion years old and your car eight years old, but neither knows it or more accurately, doesn't 'realise' it. A mountain has no awareness, as we know it, of time, but will be eroded by it and the elements such as wind which accompany it. As it erodes over millions of years, its molecules and therefore atoms are swept away. These are recycled and reused.

Most matter with which we are familiar within the Universe, with a few possible exceptions, including black holes, sub-atomic particles and possibly dark matter, is made of atoms; black holes may break atoms down into sub-atomic particles, but science is yet to discover the workings of a black hole beyond its event horizon, and deep within.

It is likely that time can only be realised by living things. Trillions of human ova and sperm that never conjoined to create life. Trillions of possible opportunities for life lost to eternity, plus the generations that could have been produced by them, had unlimited resources existed. Time had no familiarity with 'potential human beings, that never were'. The atoms that would have made up the molecules of their bodies that could have been, now populate various other objects on our Earth.

During sleep, individuals often feel that time passes more swiftly due to a temporary loss of consciousness. When under general anaesthetic, time passes even more quickly, a greater degree of consciousness is lost. Having had five operations under general anaesthetic of between one and five hours, I can witness that if I had been informed I had been unconscious for a few seconds, I would have believed it.

An induced coma can extend for weeks, months, or even years. When a person wakes up, they might not be aware of how long they were unconscious. However, the human brain often perceives this duration as significantly shorter than it truly is. Time and consciousness have

CHAPTER 8

a close relationship; it is reasonable to assume that, no consciousness equals no time.

On average, a human spends about eight hours sleeping every night, and this duration feels shorter than eight hours of wakefulness to the conscious mind. It is reasonable to conclude that when consciousness ceases at death, all time will pass in an instant or that time effectively stops, which amounts to the same thing. Conversely, when a human is in pain or being tortured, time appears to slow and take longer to pass.

It is also reasonable to conclude that although time exists, the quantification of time only exists for the living and their realisation. When realisation ends, so, it appears, does time. As discussed elsewhere, personal eternity is highly likely to be the length of our realisation, life plus womb. Unlike other animals, the human brain enables us to conceptualise time before we existed and after we die, ironic as our time alive would appear to be our eternity.

In conclusion, it seems reasonable to conclude that 'past' time directly relates to levels of consciousness. Lower consciousness equals faster time, no consciousness in death equals no time, or all time passes instantly.

Time contraction within life is a common human experience – the older we get the faster time appears to pass. This is a trick of the length of realisation, at two years old one year is 50 per cent of a human's realisation outside the womb and appears to be a long time, relative to a realisation lived, whereas a year at 50 years old is two per cent of a life lived to that date. This gives rise to a common expression amongst older humans, 'isn't this year passing quickly'. This is the effect of relativity of recent time passed against life and realisation lived.

The exploits of Alexander the Great, Genghis Khan, Julius Caesar, Napoleon and many other icons of history are familiar to many humans, yet they existed centuries before those who are, today, aware of their deeds existed. Humans documented their achievements throughout history, and as victors of numerous battles; they may not have always reported their actions accurately. Nevertheless, their accomplishments have inspired people to reflect on the past, before they themselves lived.

All humans expect to die at a time in the future, many leave instructions through wills regarding the distribution of accumulated wealth during their lifetime. After death, individuals maintain a temporary legal status even though they no longer exist. This ability to conceptualise time either side of life is an ability unique to humans.

At the end of our lives, when consciousness ends, it is reasonable to assume realisation ends. Realisation ultimately represents consciousness. A realisation we have had since before birth, yet what dies on the last day is far from what the human was physically at birth or ten years before death. Every part of us has been replicated and replaced on an ongoing basis since before birth.

CHAPTER 8

Most of our body cells are replaced over months and years depending on their function. Molecules and atoms once part of our very being and existence now spread around the Earth. Atoms that were part of us, part of our ancestors and those never to be part of humans that never existed, through unfertilised ova or failed sperm, now dispersed somewhere on our planet. Some used in other life but others, part of inanimate objects.

In Western society, more of a human who lives to be 70 years of age has been emptied out of a Hoover bag than dies at the end of life, many times this amount, flushed down a toilet in the form of dead cells, during a human lifetime.

Time has close links to every life on Earth, it defines length of survival, for humans it controls and defines our lives, when to eat, sleep, work and exist. Surely a candidate for a god, although easily manipulated by mass and gravity.

Gravity

Gravity ranks as one of the four strongest forces recognised in our Universe; it appears weak when compared to the strong nuclear force. However, if influence is compared, gravity is rightfully viewed as an awesome power within the Universe.

It can literally make time stand still. We know that strong nuclear power is a localised power, isolated and likely to burn out. Gravity has had huge, widespread influence across the Universe almost since it began. During a supernova, gravity can crush a mass of iron the size of Mount Everest to the size of a grain of sand.

Gravity pulls mass together, resulting in a denser mass and an increase in gravitational force. It plays a crucial role in the birth and death of stars, the formation of planets, and maintaining the stability of galaxies and solar systems. If gravity ceased to exist, Earth drift away from the Sun, and the Moon separate from Earth, making it impossible for multicellular life to survive. Jupiter could not have been our protector from mighty planet-destroying asteroids and meteors.

Isaac Newton is credited with discovering gravity. He stated that mass is attracted to mass. The smallest atoms have mass and therefore a gravitational influence, which enables attraction to greater mass. The accumulation of huge amounts of mass eventually resulted in the formation of stars and planets, everything we see around us.

An increase in mass within a single 'bundle' intensifies the gravitational 'pull' of that bundle. The Moon's mass is 1.2% of the Earth's, which accounts for its gravity being one sixth as strong as Earth's.

Although the Andromeda galaxy is 2.5 million light years away, it still has a gravitational relationship, or at least dark matter inspired relationship which will become a stronger gravitational relationship with our galaxy over time, the two will eventually join and form a single

CHAPTER 8

galaxy. Combined forces and influence over nearly 1.5 sextillion miles of space, 1×10^{21}.

Gravity, mass, and spacetime are interconnected in our Universe, with each element influencing the others. Gravity would be weak without mass, mass would not exist as we know it, if not for gravity. Without gravity, mass may only consist of sub-atomic particles. A greater mass results in a more significant slowdown of time; therefore, time moves faster on the Moon but slower on Jupiter or around the Sun due to varying masses.

This suggests that before the Big Bang, there was no time, if mass and gravity did not exist.

Gravity, an all-pervading force, that keeps the Universe together and functioning. Certainly, an indirect giver of life and a candidate for God.

THE LESSER POWERED KNOWN GODS

Other contenders for the title of Gods of the Universe might be lurking in hiding, waiting to be discovered fully. Scientists recognise the existence and effect of dark matter and dark energy based on their influence on the Universe's behaviour; but as already discussed they understand little about these phenomena. The same could be said of black holes and gravitational waves 100 years ago.

Science and especially study of our Universe is a dynamic study which has accelerated incredibly over the past 30 years – it is still gaining speed. I hope, after my death, some like-minded person rewrites this book or at least updates it with the latest news from science.

I am guessing that dark matter and dark energy will have their own paragraphs and will be candidates for God. As stated, initial indications are that dark energy is responsible for accelerating the movement of galaxies away from one another; if so, a strong candidate for God-like powers. I am not precious about these writings and will promise not to subject anyone to torture or death in my name, or that of electromagnetism, should they choose to disagree with me.

Many will say, we can't totally discount some kind of intelligent, all-powerful being. I have disregarded any god as described by humans in the mind of humans, in my own mind. I find the arrogance in declaring a god with such a wide remit as the Universe, to be claimed by humans in their own image as hopeful, at best, certainly arrogant. Also, being the judge and jury on good, evil and morality when neither existed for 13.8 billion years, when modern humans adopted them, as starkly amusing given some of the morality displayed by members of the church and governments.

One can reconcile all things, including scientific discoveries with God, by stating through religious commentaries that 'he' created everything. However, the concept of God and the scriptures recording 'his' actions, motives and will, were articulated by ancient peoples with little of the knowledge held by science today. When that information is challenged, then the existence of God is challenged, as it is through unsubstantiated information, not knowledge, that the idea of God has been conceptualised and propagated by humans.

It is now proven, beyond reasonable doubt, that religious writers were highly inaccurate in explaining creation, the role and significance of humans, the significance of the Earth within our solar system and Universe. Why should the unproven or unprovable parts of their testament remain influential to intellectual human thinking?

CANDIDATES FOR GODS OF OUR GALAXY - THE MILKY WAY

This is undisputed. Sagittarius A* stands out as the uncontested champion. The mass falls between 4.3 and 4.6 million times that of the sun. Its active and destructive life cycle has destroyed, possibly billions of stars and planets, or at least absorbed smaller black holes, that performed those deeds. Its recycling of matter by going quasar in the past has provided materials and building blocks of our renewable galaxy, including our own solar system.

Sagittarius A star is highly likely to swallow what remains of our dead solar system at a distant time; all that we are will one day be part of it, unless by some energy surge, a quasar strike for example, we are propelled into deep space, destroyed or homeless with no galaxy with which to belong.

Relative to our galaxy, Sag A star is indeed mighty, but compared with ultra-massive black holes, in other galaxies, insignificant, a light snack if they met. Although the undisputed overlord of our galaxy, there are still many elements outside our solar system and contained within our galaxy besides Sagittarius A star, that could be responsible for the end of our Solar System before its natural time, in about five billion years.

Gravitational waves and gamma rays from exploding binary star systems, neutron stars, hypernova and others. All potentially life giving and life taking forces, yet minor candidates for God on a galactical scale when compared with our supermassive black hole.

CANDIDATES FOR GODS OF OUR SOLAR SYSTEM

The Sun

At 300,000 times the mass of the Earth and initially loaded with ten billion years of fuel, this allows it to burn hot enough to provide sufficient heat to sustain life, to reach Earth. Hot enough to support life but not hot enough to destroy it on Earth, at its present distance and

CHAPTER 8

level of activity, and with Earth's protective atmosphere, as it destroyed life on planet Mars.

Life on Earth is dependent on the Sun, immediately and in the long term. When it first formed, the Sun appeared white with a blue tint. Over time, it changed to white, shifted to a yellow shade, and eventually deepened into the colour we see today. In the future, its colour will change to orange, and as its fuel dwindles, it will shift to red and begin consuming planets. Eventually, it will expand to a red giant, planet-consuming star, then to a white dwarf star, slowly losing all energy, converting to a black dwarf.

Smaller than many stars in our galaxy and therefore longer lived, the Sun is approximately five billion years old, with about five billion more to come, having given rise to, and sustained, complex and intelligent life on at least one planet. It has punched above its weight, ironically if it had been considerably larger it may have died before multicellular life had sufficient time to evolve.

Life giving, life sustaining, life taking, many religions have rightly viewed the Sun as a god. A far more logical conclusion than something invisible and inaccessible. Hail the God of our solar system.

Jupiter

At first view a hostile killer. Gravity that would crush humans and their spacecraft. Winds of 900 miles per hour, temperatures of -110 Celsius, no hope of humans ever approaching its surface. Yet it is a life giver – without Jupiter our Earth may be as desolate and lifeless as Mercury, Venus and Mars.

We have previously mentioned that our solar system is a hostile environment. Debris remaining from the solar system's formation combined with fast-moving visitors from beyond our solar system slice through space at remarkable speeds. Jupiter's gravity attracts large mass towards it, additionally it holds the asteroid belt in place between itself and Mars. This contains large objects up to more than one per cent of the Moon's mass.

The event that caused the extinction 66 million years ago likely resulted in the loss of 76% of all life on Earth, showcasing the destructive power of space debris. An asteroid between 6-7 miles in size hit the Earth at a speed of nearly 45,000 miles an hour, 72,000km per hour, destroyed not only the dinosaurs, but millions of species of sea life, animal land dwellers and plant life.

Of more concern is the presence of larger asteroids in our solar system compared to the one that triggered this extinction. Shortly after the Earth was formed, it was regularly subjected to bombardments of asteroids tens of miles wide and up the size of a small planet at shortly after its creation.

Although it did not prevent the initial onslaught, Jupiter has acted as protector of Earth since shortly after

CHAPTER 8

the creation of our solar system, its immense gravity attracting asteroids, meteors and other deadly life-ending objects away from Earth's path. The downside is that its powerful gravity can slingshot debris towards our planet, yet its contribution to life on Earth remains a massive net plus.

An evil life-giver who surely deserves a nomination for a god in our solar system, although it will fall short, more a 'Lord Protector', God's representative in our solar system.

CANDIDATES FOR GODS OF EARTH

The Moon

The Moon plays a crucial role in sustaining complex life on Earth. The gravitational pull it exerts on our seas and oceans generates tides that oxygenate the waters and foster an ecosystem capable of supporting multicellular life. This enabled the evolution of sea-going animals, some of which eventually became land dwelling animals, our distant ancestors.

Each year, the moon drifts away from Earth's gravitational pull by approximately 3.8 centimetres, which indicates that its influence on Earth's oceans will gradually diminish and may ultimately disappear. That event will undoubtedly pose a significant threat to long term life, especially multicellular life, on Earth.

Moreover, the Moon stabilises Earth's tilt, which in turn regulates the climate and seasons that allow life to persist and flourish. If not for this stabilising effect, planet Earth wouldn't have the diverse life forms we see today. Humans almost certainly being a 'never created' casualty of that absence.

A short testament to a nevertheless essential component in the evolution and sustaining of life on Earth.

The Oceans and Seas

The Moon drives the tides of our oceans and seas, serving as the cradle of life on Earth. Saltwater sustained life until organisms evolved to breathe air on land, transforming the existence of every creature that flies or walks around you.

The Sun's process of evaporating saltwater into freshwater clouds made a significant contribution. For nearly two million years, heavy rainfall shaped an environment where trees and other life forms that produce oxygen transformed the Earth's atmosphere, ultimately generating enough oxygen to support enormous creatures like dinosaurs, which stand as the largest land-dwelling creatures, since life on Earth began.

CHAPTER 8

Today, there is less oxygen in Earth's atmosphere, probably not enough to support land goers of such size, but enough to support humans and an abundance of flora and fauna. The seas and oceans, linked and mainly unbroken mass of saltwater, covering 65% of our planet, life giving, life taking. Surely a candidate for a God on Earth.

Nature, Natural Selection and Evolution

The three are indistinguishable, similar to the father, son and holy ghost/ spirit in Christian teachings. They embody the journey we have all benefited from and exist because of. From the 300 million years struggle for life to get a foothold on Earth, through 3.8 billion years' sustainable life, in firstly single cell creatures, then the development of DNA, mitochondria and photosynthesis in cells. The adaption of life to its changing environments in order to ensure survival, making it stronger, fitter, faster and more self-sufficient. Some animals become self-sufficient at birth, while others depend on their parents yet possess more advanced brains that mature over several years.

All animals today sit on Darwin's imaginary tree of life, on the extremes of ever growing, imaginary branches. Humans may get the impression that they sit at the top of an evolutionary ladder.

Every animal, plant, fungus, and microbe exists at the peak of its unique 3.8 billion years evolutionary journey. Contributing to the ongoing development of its species, but only for a short period. Humans may have the edge when it comes to brain power, but the cheetah has speed, the hunting dog endurance and the giraffe a longer neck. All adaptions by nature through the bedfellows of natural selection and evolution.

Natural selection has shaped life to respond to the constant challenges posed by our dynamic, vibrant, lush, sometimes icy, and at other times sweltering planet.

Surely a force that is a candidate for God on Earth.

Humankind

Since the inception of civilisation, humankind has established kingdoms, empires and armies, and developed nuclear weapons, thus harnessing immense power. Certain individuals have proclaimed themselves as divine beings on Earth. The power of life and death over billions of people and animals.

Humans have utilised the later part of the cognitive revolution to secure food supplies without relying on foraging, leading them to live vastly different lives from those of their ancestors. The only other animals that hold this distinction are either domesticated or have lost their natural habitats because of human activities.

Time disqualifies humankind from being considered a God. Gods can't be that temporary unless they are in

CHAPTER 8

the minds of humans. Humankind has only had these high reasoning powers for a relatively short time and is possibly destined to destroy itself, be destroyed by an extinction event or develop into another primate with a yet further developed brain. The status quo is not an option for humans. Humans are creators and destroyers, giving life and taking it away, but with brains and existence, too vulnerable, for such elevated status.

Bacteria and Microbes

Bacteria with godlike status. Ridiculous! Maybe.

As discussed earlier, bacteria are in us, with us, every moment of our lives. We can't live without them, yet life existed for billions of years, without conceptualising a God. Unicellular life dominated Earth for two billion years and will potentially live for more than 80% of the life of the Sun. Homo sapiens and their gods will live for a fraction of that time. Early human gods, and their followers, are testimony to that.

Bacteria enrich soil, decompose waste, act as nature's recyclers, ensuring conditions are suitable for the next generations of life, by using death of previous generations and waste products of current generations. They are key to the functioning in the Gaia Principle developed by Lynn Margulis and James Lovelock. This proposes that 'all organisms and their inorganic surroundings are closely integrated to form a single self-regulating complex system, maintaining the conditions of life on the planet'.

'The Gaia Principle posits that the Earth is a self-regulating complex system involving the biosphere, the atmosphere, the hydrosphere and the pedosphere, tightly coupled as an evolving system. The theory sustains that this system as a whole, called Gaia, seeks physical and chemical environment optimal to contemporary life'. Harvard University

Humans and all other animals, plants and fungi serve bacteria, we provide perfect environments, enriched with moisture, oxygen, food and warmth. It is as if bacteria designed animals, plants and fungi for their own use. Humans have learned to kill some varieties, yet they develop and update resistance.

Humans can't kill all varieties as they would kill themselves. We live and die, bacteria split in two, using the similar genetic coding as their ancestors. Despite incalculable deaths of microbes over a short period of time, often measured in minutes, an incalculable number remain alive, testimony to their powers of survival.

Prior to multicellular life cohabiting Earth, microbes had a much tougher existence, conditions were exposed and harsh, travel was by water or airborne. Animals transport them around the globe, humans on aircraft, birds migrating, whales following food, bacteria and

CHAPTER 8

other microbes have a fruitful existence thanks to the multicellular animals they created.

We know DNA has an unconscious 'intelligence' and a 'learning capability' through changes in chemical structure and coding. Intelligence within a different dimension to that with which humans are familiar. Nevertheless, DNA coding has the inbuilt capability to adjust and adapt bodies, minds, functions and processes of the animal, plant and fungi kingdoms.

It can influence a fertilised animal egg and within a short time, place the organs, blood vessels, muscles, tendons etc in exactly the correct position for the developing embryo. It can pre-programme the behaviours of living things, for whole life.

DNA is not conscious within the dimensions of consciousness that humans are familiar with, yet its powers of adaption and therefore learning are beyond the powers of the human mind. DNA within somatic cells solve a multi-trillion pieces jigsaw with every human birth.

Originally and exclusively within microbes, DNA for two billion years experienced a fight to constantly adapt unicellular organisms, to survive and thrive within severely harsh environments, which could not sustain the smallest of multicellular life.

Bacteria and other microbes dominated Earth for two billion years, with little visible progress, then the environment changed radically, enabling multicellular animals to thrive. This also enabled microbes to thrive in greater dimensions and vastly increased numbers, as they were hosted by larger organisms.

Could this development have been the nonconscious but 'intelligent' strategy of microbe DNA to create a better long-term future for themselves, now that conditions were suitable for larger organisms? There are billons of times more microbes living on Earth than animals, plants or fungi. Microbes clearly benefited from the expansion of multicellular life, arguably more than multicellular life itself, given the additional numbers created. One animal life, giving life to billions or trillions of microbes, dependent on size. Microbes are clearly the survival experts and victors.

Whatever life is like in other parts of the Universe, we can be reasonably certain that it started from microbes, unicellular life. What it became or becomes, on those far off worlds, will be determined by environment. A plentiful supply of metabolisers (oxygen on Earth, for example) and sources of energy will lay the foundations for multicellular organisms, as it did on Earth.

Life was very different on Earth two billion years ago, 800 million years ago and 150 million years ago due to oxygen being at varying levels. Should life be found on other planets it is expected to be very different from contemporary life on Earth. The mix of Gaia levels, temperatures, environments, levels of water, carbon,

CHAPTER 8

nitrogen, oxygen and many other factors will determine unicellular creatures' body form, they in turn, with the aid of natural selection, would determine multicellular life forms. A one percent change in metabolisers would change life and DNA to some degree.

When we take eyewitness accounts and science fiction's conclusions of what form intelligent life, from other worlds appears, when visiting Earth, we need to consider evolution. Small white people with large eyes suggests a dark environment, with less nutrition than on Earth. Large eyes would only develop in a dark environment and white skin confirms a lack of sunlight.

The variety of environments, in terms of water type and content, atmospheric mix, sea and land ratio, rainfall, proximity to a star, temperature levels and variation, light levels, length of days, existence of seasons and thousands more inputs would determine alien life form appearance. Remember, high oxygen levels enabled large dinosaurs to exist, highly unlikely on modern day Earth with between 4-14% less oxygen than the time of history's largest land dwellers.

Bacteria, microbes or similar unicellular organisms, are likely to dominate the Universe, past, present and future. They can inhabit planets and moons where multicellular life would perish. They exist much sooner after the creation of a solar system, and may still exist on the outer planets and moons shortly before their star goes supernovae or implodes.

Unicellular life is, therefore, likely to inhabit millions of trillions of planets and moons across the Universe, that multicellular life can't, without inhabiting one per cent of planets and moons. For two billion years our own planet was in such a state. We can be confident that where multicellular and intelligent life persists, unicellular life was the driving force and foundation of that life.

A life creator, a life preserver of life, a taker of life, likely to exist throughout the Universe. Some good, some evil, but only in the views of humankind. We are the product of their DNA, they are our Genesis. Bacteria and microbes godlike… ridiculous… maybe not.

SUMMARY

With the exception of humankind, all of the above candidates for God have played a vital role in producing multicellular life on Earth, and therefore us. It appears that those humans who practised polytheism had justification as a number of different forces and powers were responsible for creating and sustaining life. Powers with no intelligence or consciousness, using the human definitions, yet highly influential on the existence of life. No consciousness, yet key to everything; human defined consciousness or intelligence is maybe not a requirement for a god.

For example, DNA has unconscious intelligence, it has influence through its varied and changing chemical

CHAPTER 8

construct, is key to life and evolution. Different dimensions of intelligence, held in the brain through which humankind has developed, discovered and invented an array of wonders. Compared to the powers of the Universe, these achievements are minor, as they are when compared to how natural selection and DNA has determined all life, its survival up to and including the incredible yet vulnerable and superstitious human brain.

*

References, sources and recommended reading or research, Chapter 8

How the Universe Works - *Pioneer Productions / Discovery Channel, 1.2,1.8,4.5,4.6,4.7.4,10.3 & 11.3*

Gravity - *NASA spaceplace*

Black Holes - *NASAscience.gov*

Quasars: Brightest objects in the Universe - *Space.co, Keith Cooper*

Electromagnetism - *accessscience.com*

Gravitational Time Dilation - *Dr Chris S Baird*

Isaac Newton, Biography, Facts, Discoveries, Laws and Inventions - *Britannica.com*

Is it true that Jupiter protects Earth - *Deborah Byrd, EarthSky*

Strong Nuclear Force - *Richard Webb, New Scientist*

Was God a Bacterium - *Daniel Witt, evolutionnews.org*

CHAPTER 9

What Lies Beyond

Two questions have fuelled my curiosity, yet I will never know the answers. Firstly, what is the long-term future for humanity, if it survives? What will humanity become, what is the next stop on the continuing journey of evolution? Yet the past tells a lot about the highly likely, if not certain, outcomes.

There are two apparent certainties: the first that the Earth and all in our solar system will come to an end; the second certainty is that Homo sapiens, long before that time, will continue to evolve, but more likely intellect wise than physical deviation, and at a faster rate, possibly a vastly faster rate. Everything else related to our future has guesswork attached. In the short term, how AI will affect humanity, will biological weapons of mass destruction be released with devastating consequence in our global world, with global warming already destroying our eco-system.

Covid showed us how fast disease can spread in a world where over 100,000 commercial flights take place, each day across our planet. These outcomes are, however, all guesswork. The diversification of humanity into co-existing species of humans and the eventual destruction of the Earth, both appear inevitable.

Numerous potential events exist, far too many to cover in a brief book. My focus will remain on the two previously highlighted aspects while exploring, in a hypothetical sense, how humans and life could potentially face extinction and how religion may fare in future. This list maybe far from complete. Additionally, I will examine how the balance has changed with the emergence of modern humans.

My interest in evolution has enabled me to look back on 3.8 billion years of life and its struggle to adapt. Yet, in less than 200 years, humankind has changed the Earth to such an extent that Homo sapiens from 300,000 years ago up to 200 years ago, would not recognise it. More change to Earth and humankind in 200 years, than 300,000 years of Homo sapiens that preceded it.

This evidence overwhelmingly shows that humankind has accelerated its brain power along with the discoveries and inventions that enrich our daily lives. Over time, human progress seems to accelerate at an unprecedented rate compared to earlier periods. I have examined the past and attempted to forecast the future for the next few hundred, up to two thousand years.

CHAPTER 9

I then address the long term, neither in exceptional detail – this book is to give some additional knowledge, educated guesses and supported theories, but more, to fuel the interest of the reader, so they will want to explore further.

Humans' journey has been a mix of many discoveries and inventions that have signalled progress, others that have signalled danger. For example, we can cure the previously incurable and have developed the capability to destroy humankind through weaponry, laboratory developed microbes and atmospheric emissions.

Survival has been the main driver of life, including humans, since the first microbes. About 6000 years ago humans developed a challenger to this driver in gods and organised religion, which became all-consuming for many. Many have given their lives on behalf of gods, believing that swapping life for various forms of utopian eternity. This fascinates me, in a similar way that a relatively small amount of people turned the minds of millions in Nazi Germany – added to my interest in the human brain, it is a subject I return to often. Lives laid down for a belief, feeling, an emotion, it's incredible. I believe an early and unnecessary walk into oblivion.

The second question is, how will it all end for humanity, life on Earth and our solar system? These three endings may be billions of years apart, but all three are coming. I will cover this first.

DOGS THAT EAT GODS

DESTINY, CERTAIN AND IRREVERSIBLE

'We only need to look up, to know the eventual fate of our solar system'. Michelle Thaller, Astronomer, How the Universe Works, Discovery+

In the past 30 years, astonishing discoveries about space have provided insights into the Universe, encompassing the eventual fate of our Solar System and life on Earth. The almost infinite number of stars, solar systems, in the Universe and our ability to observe trillions of these, over time.

Telescopes in space, like Hubble and James Webb, observe stars with mass comparable to our own in all stages of their life cycles, from newly formed to giant red stars nearing the end of their approximately 10 billion years lifespan. Our star has reached about the midpoint of its lifecycle, and we therefore have an approximate estimate of when it will perish.

Solar systems form and disintegrate on a conveyor belt-like cycle across extensive timescales, of millions, billions or even trillions of years, influenced by the star's size and its interactions with other stars and celestial entities, such as black holes and quasars. Humans feel comfort in believing our planet and solar system are special, yet it is highly likely similar stories have been played out across our galaxy many times, across our Universe, many

CHAPTER 9

trillions of times. Everything is destined to end, humans within a few hundred thousand years possibly, animals within hundreds of million years and bacteria within a few billion years.

Scientists estimate that as many as 85% of stars exist in binary systems or contain multiple other stars within their systems, amounting to three or more. Their eventual deaths may be harder to predict with larger stars dying earlier and therefore likely to cause damage and change to their smaller co-star(s) in close proximity.

As we know, length of life of stars is related to size, exceptionally large stars may die after only a few million years, a mere fraction, less than 1/1000th, of the expected life of our own Sun.

The future might not be as challenging to predict, as we believe. The elements that make up our bodies, our planet and our solar system were formed in dying stars, in other parts of our own Galaxy and possibly the wider Universe. We understand that our Star will eventually die, but it will not be in vain. The elements that formed within it and within us are likely to contribute to the creation of future stars, planets, and perhaps even life itself.

The long-term fate of Earth and life on it, is easier to predict. If the Earth is fortunate enough to avoid a terminal collision, a quasar energy bolt or gravitational waves from two neutron stars colliding in relatively close proximity etc, then the Sun will eventually consume it. Before that time, complex life will face extinction hundreds of millions of years in advance due to the heat from an expanded Sun scorching this planet.

We have explored the powers of our Galaxy and the wider Universe earlier in this book, we understand that they are life giving but also destroyers of life. What threat do these pose to life on Earth? Many different threats will face Earth and life in future – I will cover some of the main ones which have occurred to me whilst researching this book.

AN ASTEROID OR METEOR

The Earth has experience – since its formation it has likely collided with trillions of objects ranging from smaller than a golf ball to the size of a small planet. The asteroid of 66 million years ago which made dinosaurs extinct wrought devastation on life, yet Earth and life recovered.

Our Sun is barely halfway through its lifecycle and should humans and most other large mammals be made extinct, there is plentiful time left for evolution to produce a species as intelligent or more intelligent than humans, and maybe less destructive, providing multicellular life remains as a starting point. Evolution will start again with the remaining gene pool. In a

CHAPTER 9

million years plastic and fossils would be the only sign humans ever existed.

Interim danger to complex life on Earth from space is more likely to come from an asteroid or meteor than more destructive powers light years away. Scientists believe our Earth had numerous large collisions in its early life and that the Moon was formed by one of the more cataclysmic events.

The asteroid that struck Earth 66 million years ago measured about six to seven miles across (10-11km) and led to the extinction of three quarters of life on the planet. Much larger objects wander across our solar system travelling at high speeds that could be far more disastrous for life on Earth, should they meet.

It is a danger that is being addressed. Humans do have technology to intercept danger from space by detonating explosives to destroy or more likely deflect the course of solid objects headed our way. However, this is not a failsafe process − large objects travelling at over 70,000km per hour could prove difficult to divert or destroy.

The danger has significantly decreased compared to just a few decades ago due to the human brain adapting to meet the challenges. Observing our solar system allows us to provide sufficient warning, calculate trajectories and impacts, and develop intercept measures. The human brain has minimised certain risks but has also heightened other types of risks.

THE MOON MOVING AWAY FROM EARTH

We know the Moon is moving away from the Earth at a few centimetres a year. The slowing of the Earth's rotation and the lengthening of days is diminishing the Earth's gravitational pull. The Moon has been key to life in Earth's seas, which became life on land, by creating tidal movement and bringing oxygen to life in our oceans. When that influence is weakened sufficiently, life in our oceans is likely to die. The effect on the food cycle of life on land could be devastating, potentially terminal.

The weakening relationship between Earth and Moon will inevitably disrupt life on Earth, natural selection will be challenged to adapt life to new environments and is likely to, eventually, be defeated. However, other events are likely to destroy human life on Earth before the loss of the Moon's influence.

LOSS OF EARTH'S MAGNETIC FIELD

We know that planet Mars had rivers, oceans, an atmosphere and probably life, at least at unicellular level. Earlier, I mentioned that Mars likely lost its electromagnetic field due to the cooling of its core. This event led to the destruction of the planet's atmosphere, exposing its surface to harsh solar winds and resulting in the red, barren planet we see today.

Fortunately, Mars measures just over half the size of Earth and possesses only about 11% of Earth's mass. The cooling of Earth's core will take a significantly longer time period, leading to prolonged electromagnetic protection from its poles. It will eventually cool down, but another event is probably more likely to wipe out humanity long before that happens.

SOLAR BLASTS

From the point of view of life on Earth we have been fortunate to have such a stable star to call our Sun. The Earth's atmosphere protects life on Earth when our Sun becomes active and releases solar flares. This should continue for hundreds of millions of years; the Sun will, though, become more unstable with age.

If Earth's atmosphere is damaged, but not destroyed, by another major event such gravitational waves, the after-effects of a neighbouring star going supernova, a quasar etc, then Earth and its life could be exposed to more atmospheric damage. This would make Earth far more vulnerable to the irregular activity of the Sun.

Damage to the ozone layer of the Earth's atmosphere has, for example, increased the level of damage to humans from harmful solar rays. This is comparatively small scale compared with the effect serious damage to Earth's atmosphere could cause. Repeated damage to the Earth's protective atmosphere may pose a realistic threat to life on Earth in future.

Long term, the expansion of the Sun will claim the Earth, a gradual process, therefore difficult for scientists to predict; however, complex life is unlikely to exist around a billion plus years from now.

GRAVITATIONAL WAVES

A threat remains invisible and undetectable until it strikes.

In 1893, Oliver Heaviside first proposed the concept of gravitational waves, followed by Henri Poincaré in 1905. Albert Einstein also predicted their existence in his 1916 'General Theory of Relativity', as ripples in spacetime. The first direct observation of gravitational waves was in 2015, when energy generated by two colliding black holes was received by detectors in Washington State and Louisiana. The importance to Science was so significant that the Nobel Prize for Physics was awarded to Weiss, Thorne and Barish for their discovery.

Sources of gravitational waves include binary star systems composed of white dwarfs, neutron stars and black holes, plus events such as supernovae. The level of energy

CHAPTER 9

released and the proximity of the event, determines the destructive capability.

A collision between two neutron stars situated near our solar system could significantly harm our planet and might lead to an extinction event that affects humans, though it is highly improbable that this scenario would be the ultimate cause of Earth's destruction. The power gravitational waves possess, and the range of their influence, a reminder of how small and insignificant our planet is.

Scientists calculated that the gravitational waves detected in 2015 originated from a collision between two black holes that occurred 1.3 billion years ago. The gravitational waves had been travelling to Earth at the speed of light for over a billion years before the first dinosaurs existed on Earth. An unlikely way for humanity to end but it can't be discounted.

QUASARS

Quasars, the Universe's 'bad boys'. Black holes that accumulate matter around their accretion disk, drive it to their poles at incredible temperatures, then jettison massive energy and matter light years across space, destroying anything in its path.

As mentioned earlier, more release of energy in one second than our Sun will release in its entire ten billion years lifetime. Should our solar system be in the path of energy released by quasar and close enough, it would be destroyed, possibly vapourised in minutes. Again, an unlikely event compared to the risk levels of other cataclysmic events.

An unlikely force to destroy Earth and life on it; however, the recent discovery that up to 100 million black holes could exist in the Milky Way, a threat that cannot be dismissed.

BLACK HOLES

Recent discoveries by scientists show that black holes are more common than previously thought. Gamma rays pulsating through space are the 'signature', signifying the birth of a black hole. A new black hole is detected on average once a day, indicating many trillions of black holes throughout the Universe, the great majority too far away to detect. The vast majority of these will be less than supermassive, likely destined, in the future, to become part of a supermassive or ultra-massive black hole, as gravity unites them.

A quasar, a highly active black hole, yet black holes continue be high risk to stars and planets even when not in quasar mode. Sagittarius A star at the centre of our Galaxy is of little danger to Earth, for the foreseeable future. A distance of 26,000 light years should ensure that our solar system is destroyed by other forces long before being destroyed by our supermassive blackhole.

CHAPTER 9

Stars drawn into close proximity will not be as fortunate. Black holes exist closer to our solar system, smaller than Sag A star yet can still be hundreds or thousands of times the mass of the sun. Although only around 50 black holes have been detected in the Milky Way, scientists believe this number could be up to 100 million in our galaxy alone – explained by the death of large stars, formation of black holes and the likely absorption of stars during the lifetime of the Milky Way.

These figures clearly increase the likelihood of a black hole destroying our solar system in the distant future. A large star going supernova within a relatively close proximity of Earth would therefore create the double risk of Earth being destroyed by the 'blast' or later being drawn towards a newly created, by extreme gravitational force, of the black hole.

It is unlikely our solar system will be destroyed by a black hole before humans become extinct, yet highly likely that is where our long deceased, incinerated Earth and spent Sun possibly having been recycled several times, will find their final resting place. Black holes are the 'Hoovers' of a Galaxy, gorging on stars, planets and gases, that get too close.

SUPERNOVA

There are different kinds of supernovae, yet any large enough and close enough they could cause extinction events on Earth.

Scientists have discovered Iron 60 when exploring Earth's oceans; this element forms exclusively in a supernova. It is found in all oceans of the Earth and is at a depth within ocean floors, consistent with it arriving on Earth 2-2.6 million years ago. This means that a supernova occurred close enough to do significant damage to Earth. Although not an extinction event, it is estimated that a third of marine life died out as a result. The casualties were mainly 'shallow depth' dwellers with deeper water species being less affected.

The last recorded supernova in our galaxy was recorded 400 years ago and was too far away to cause any damage. Danger would be relative to size and proximity. A large star 20-100 times the mass of our Sun may cause damage at a distance of 150 light years, a medium-sized star may cause damage at 30 to 50 light years. An enormous star exploding at that distance could potentially lead to extinction-level destruction.

Scientists may eventually observe a potential supernova large enough and close enough to threaten life on Earth, within the coming decades, centuries or millennia; it will occur at some future time, as it did around 2.6 million years ago.

A span of two to 2.6 million years represents a relatively recent chapter in evolutionary history of Earth, humankind Homo habilis roamed the Earth, indicating a significant threat to the Earth and especially its atmosphere. If Earth loses its atmosphere, it loses all

CHAPTER 9

multicellular life and its surface water. Potentially the worst post-Cambrian extinction event of all, beating the devastation caused by the most destructive event around a quarter of a billion years ago.

A significant number of scientists believe that the Devonian Extinction, about 360 million years ago, was caused by a supernova approximately 65 light years from Earth. This is believed to have decimated 97% of water-borne vertebrates. A repeat of similar would almost certainly make humanity extinct – any more powerful and life on Earth could be threatened.

Approximately 60,000 stars are visible within 100 light years of Earth; this represents 0.00003% of the stars in our Galaxy. A supernova does not need to cause a mass extinction to threaten humanity, damage to the Earth's atmosphere could allow damaging cosmic rays to access life on Earth. This could damage DNA, causing widespread cancers.

Supernova events pose a greater danger compared to other external influences on our solar system, but they are probably more of a long-term threat. The nearest star observable and likely to go supernova in the foreseeable future is 400 light years from Earth.

Supernovae, already having been responsible for life and death on Earth, a strong candidate for extinction of multicellular life on Earth.

SELF-INFLICTED EXTINCTIONS - EARTH

With the development of technology to deflect or destroy asteroids, the internal workings of Earth, whether humankind, bomb, magma or microbe has been promoted a main danger to life on Earth, in the shorter term.

The Earth has proven a great danger to life on its own surface in the past. For 300 million years, life struggled to establish a foothold until the environment of Earth changed sufficiently, then, over two billion years passed before multicellular life was able to flourish.

Eruption of volcanoes for 60,000 years, 251 million years ago, are believed to have caused, or been a major factor in the most devastating extinction evidenced by scientists in the last 500 million years. The release of carbon in huge quantities into the atmosphere caused global warming and destroyed 96% of life on Earth. Acid rain for thousands of years meant only the truly hardy, land-living animals and plants survived.

The cooling of the Earth's core since that mass extinction has led to a decrease in volcanic activity, making such events less probable in the future. Humankind is causing a similar effect through carbon emissions, and whether we can prevent it from leading to comparable disasters remains uncertain. It's ironic that while humans have decreased the danger of extinction from space, they have heightened the risk of extinction on Earth.

Humans have created additional means of mass destruction, such as nuclear weapons, biological agents, and chemical weaponry. These events probably won't annihilate all of humanity, but they could trigger a significant 'corrective' adjustment in population and potentially revert our numbers to those from the 1950s or possibly even the 1800s. If humans completely went extinct, evolution and natural selection could ironically find an opportunity to initiate a new process, beginning anew with the remaining gene pool.

The word 'corrective' is used in the context of balance of life on Earth, which has clearly become out of kilter due to the success of Homo sapiens and their negative effect on Earth's ecosystems and many species of life.

Humans pose significant threats to each other and to other species, including animals and plants, but they do not endanger the fundamental life forces on Earth – bacteria, for example. If 99.9% of life on Earth were to be eliminated, bacteria would still survive, and we would revert to conditions present two to 3.8 billion years ago, when only unicellular life existed. It would all start again with enough time remaining on the evolutionary journey, to possibly develop complex life, yet unlikely to reproduce humankind or similar, which required around 1.5 to 1.7 billion years to evolve from microbes.

By that time the Sun is likely to have expanded sufficiently to destroy complex life. The long-term future of life on Earth is not dependent on humans, it is more dependent on life we excrete. The descendants of bacteria growing in our bowels are likely to be surviving long after our descendants have reached the end of the evolutionary line.

LONG TERM - CAN HUMANS FIND A NEW HOME, BEFORE DISASTER STRIKES

The current state of humanity suggests that humans, as we know them, are unlikely to exist in their current form in 100,000 years, for one of two reasons. Firstly, we could have been made extinct by our own actions, a celestial occurrence such as an asteroid or other major event.

Alternatively – and hopefully this scenario is more likely – we will have progressed by that time into an even more complex being, as distinct from today's humans as we are from the Neanderthals of 100,000 years ago or the Homo erectus and Homo heidelbergensis of a million years ago.

What lies in wait for humanity if it is unable to colonise different worlds, is not in doubt; extinction is certain, eventually. Believing that God will intervene amounts to the same futility the monks on Holy Island experienced when they prayed for salvation from the Vikings in 793

CHAPTER 9

AD. The eventual destruction of our solar system is going to happen, the evidence is overwhelming and being played out above us. It is an inevitable and routine event.

Worlds where humans might survive away from our solar system exist many light years away. The nearest 'new' world is approximately 4.25 light years or 25 trillion miles distant. An alternative dimension of travel would need to be discovered. Current calculations mean that travelling over the distance to our nearest star, at a tenth of the speed of light, 42 years of travel, would consume more energy than humankind has ever used.

However, it is highly unlikely that a planet exists that would support life, in that nearest solar system. Humans, and what evolves from them, are unlikely to prevent a star from ageing and dying; our brains have changed the Earth, changing our solar system and travel to an alternative, would appear, for now, to be a challenge too far.

Can the human brain, or more probably a more advanced version that evolves from it, rise to meet the challenge of travelling light-years across spacetime? First, it must rise to confront the survival challenge, returning to the primal foundations that our ancestors established. Microbes have successfully addressed these challenges for 3.8 billion years. As I write human extinction seems a far more likely scenario than travelling over four light years.

*

References, sources and recommended reading or research, Chapter 9

How the Universe Works - *Pioneer Productions / Discovery Channel, 1.2, 9.8 & 10.3*

Moon facts - *NASA Science*

What are solar flares - *European Space Agency*

Gravitational Waves discovered - *New York Times*

What is Supernova - *Space.com*

CHAPTER 10

Humankind
– Still Evolving and Destined for Something Better? Then Destined to Die

---

Over the past 90 years, we have come to see significant differences in both IQ and emotional intelligence amongst humans. If left unaddressed, diversity of living and culture will expand this divide further.

Significant observable gaps between social classes within the world's economies include: differences in IQ, reading, mathematics, science and developed emotional intelligence, opportunity to learn, leisure time, nutrition and environment. When then compared to the average human in third world countries, the gap is considerable.

IQ is not a wholly or universally accepted measure of intelligence, but I will use it as a proxy; it is 'broadly commensurate' with reasoning ability, problem solving, ability in mathematics and the sciences. The foundations of economically developed and advanced economies.

Scientists state that IQs are culturally specific and therefore the validity in comparing across cultures is weakened. It is not the IQs within diverse societies or measurement techniques that I wish to highlight, it is the effect they are having in driving developed economies and poorer economies further apart. It appears clear that countries and societies with higher average IQs are more successful in achieving educational and economic success. Whether they are an effective comparator is maybe not as relevant as their status in educational and economic propulsion.

The average IQ in Japan has been recorded above 106, then a raft of advanced economies over 100 (including Germany, Taiwan, Finland etc). In the USA, which has poverty levels more than three times that of Finland, the average IQ is 97; the world average, including the high scorers is 82. Nepal scored 43-51, depending on what study is used, the country with the lowest recorded average IQ.

In 1984, James Flynn discovered a 13.8 points increase in IQ scores, between 1932 and 1978 in developed economies. A rise of about three points each decade. This trend has continued across developed economies. Similar improvements have been seen for semantic memory, general world knowledge and episodic memory, everyday events. This is likely to increase with easier access to learning and mental stimulation brought about

CHAPTER 10

by improved access to learning combined with improved learning techniques.

Research indicates a possible reversed Flynn effect, showing a decline in IQ scores in Norway, Britain, Denmark, Sweden, Finland, and Germany. This began in the 1990s. This occurs despite 15-year-olds in those countries performing well above the international average.

A study of a heat map depicting average IQ levels in countries worldwide reveals that Africa, Southern Asia, and Central and South America fall behind developed Western and Far Eastern nations in terms of intelligence quotients. Since 1990, many individuals have migrated from countries with lower average IQ scores to richer nations that have higher averages.

The six European countries named above now have populations which include between 11 and 20 per cent of people born outside their borders. All six countries have seen substantial increases in population from comparatively under-developed, non-European countries since the 1990s, many migrating from countries with poorer education systems, presumably, at the time of entry, scoring significantly lower on average IQs. Other factors such as air pollution and availability of processed foods should not be dismissed as possible reasons for this dip in recorded IQs; however, the six countries identified have relatively low levels of pollution, yet it has increased in all since Flynn's findings, as has the consumption of processed foods.

It would appear numbers of migrants may be enough to account for at least some of the retrograde IQ results as the average is forced down. Most migrants from under-developed economies are mainly over school age when arriving in their host countries, and therefore are unlikely to engage in future complex mental stimulation within education, in sufficient numbers to maintain or raise standards.

This could potentially clarify why the average IQ levels for adults have declined in certain developed countries, while children's standards continue to remain high. A valuable study could track the educational and IQ testing progress of children from migrant families, from comparatively under-developed nations, within developed economies. Improvement appears highly likely compared to the average in their parents' country of origin. Opportunity appears certain to make a difference. This may not be the case in the more distant future, if the gap is allowed to grow, to become embedded and cultural.

The widespread availability of education has certainly contributed to improvements in IQ, as it provided more individuals with access to mental stimulation and challenges. In 1880, in the UK, education became compulsory for all 5-10 years old, giving many access to education for the first time. School leaving age was gradually and continually raised. This has become the bedrock of economic success.

- 1893 – 11 years of age
- 1899 – 12

CHAPTER 10

- 1918 – 14
- 1944 – 15
- 1972 – 16
- 2013 – 17 2015 – 18

Mid-point standards have improved considerably over that period, in many Western and Far Eastern countries, whilst improving at a slower rate in less developed economies. The duration of education has lengthened while the complexity has also risen across different age groups. The rising complexity of mental stimulation is likely to enhance IQ scores and other indicators of intelligence.

The relationship of rising IQs in developed nations which introduced earlier compulsory education, strengthens the argument for complex challenge and mental stimulus being at least partly responsible for rises in IQ averages. What these nations have achieved with increased education has been incredible, again strengthening links between IQ and economic and scientific growth. Scientists may argue with how significant IQ is in measuring human intelligence, but evidence does, however, point to it being an essential ingredient in modern human economic and scientific achievements.

The UK has seen a significant increase in non-compulsory education. In 1950, approximately 3% of school leavers attended university; by 2000 this had risen to over 30%. The UK population increased by a third in that time, this suggests approximately 13 times more people attending higher education.

Reading, mathematics and science achievement broadly reflects these IQ differences. Some humans can achieve IQs exceeding 200, and when linked with high emotional intelligence, they represent an exceptional individual.

As time progresses, we will probably observe an increase in the number of people achieving this combination. If Flynn's average of three IQ points improvement, per decade, is sustained over future decades, or more likely increases, in developed economies, differences will surely accelerate to a marked and irreversible level in future. This is dependent on controlling IQ growth inhibitors such as air pollutants being reduced.

Is there a natural ceiling to human intelligence? An existing human IQ of 276 combined with human achievements over recent history, would suggest that if there is a ceiling it is high, if it exists.

The differences between IQs in developed and underdeveloped nations lie not in race, creed, or religion, but rather in wealth, opportunity, technology, and need. Developed economies depend significantly on learning to promote and advance their economic growth. However, beyond these reasons, the key lies in stretching and testing human brains; the more we challenge them, the more skilled they become.

In high-performing economies, rewards based on achievement provide an additional motivation. Natural selection, and through it the development of our 'brain

CHAPTER 10

capacity' over tens of thousands of generations, has driven humans to become changers of environments rather than relying on natural selection changing them to adapt to environments, as was the case for all living things over billions of years. A short human tenure as masters of the Earth, yet very eventful.

Learning capability has grown in humans, especially in developed countries, and we now move into the age of science and academia and away from the age of superstition. God, religion and superstition as leading human influencers, have been a great leveller within humanity for millennia; Science and Technology is having, and will have, the opposite effect. We have already discovered, evolution is about gaining advantage, not equality.

Individual IQs can fluctuate by more than 160 points across the spectrum, and those with higher scores are likely to exhibit better emotional intelligence. For context – the dictionary definition of emotional intelligence being 'the capacity to be aware of, control and express one's emotions and to handle interpersonal relationships judiciously and empathetically'. A huge evolutionary change, a further departure from primates and from some humans.

Although modern Homo sapiens have not diverged significantly from early Homo sapiens, in terms of physical DNA modifications, we should remember that there has been little pressure in the vast majority of development that DNA influences. Humans still need a stomach that digests, feet and legs that can hold balance, the Krebs cycle to produce energy plus millions of other functions.

The pressure for types of physical change through natural selection has lessened significantly as humans have conquered their environment. Shelter and especially medicine are easing pressures on evolution within some physiological characteristics. Cochran and Harpenden argue that human evolution has accelerated in the past 10,000 years, rather than slowing. They believe it is now accelerating 100 times faster than in long-term average over the six million years of our existence. Modern living in developed countries is certainly presenting mental challenges for a species which has DNA mainly developed for hunter-gathering. Recent evidence appears to label human challenges more psychological than physiological.

Evolution has ironically become driven by humans adapting environments rather than changing physically to suit environments. In short, in adapting our environments to suit ourselves, that is no less evolution than nature adapting us to suit our environments. One is changed by human intelligence, the other by traditional natural selection; both are evolution. The intelligence changes are much more difficult to measure in scientific formats such as DNA studies, yet they are becoming more obvious. We may need further measures which break out of traditional measures of species development.

CHAPTER 10

A study by Professor Beben Benyamin, University of Queensland concluded that 20-40% of a child's IQ or intelligence is due to parents or family genetics. This indicating some influence through nature, but still mainly influenced by nurture. We need to remember that diversity in education standards between third world and developed countries, for the masses, has been significant for approximately 150 years, a blink of an eye in evolutionary terms. It appears logical that further diversity is inevitable over longer time frames if differences in opportunity continue to be sustained.

The gene / protein that Benyamin identified is called FNBP1L responsible for this difference, this concluded that 60-80% of intelligence was not due to genetics. I conclude that if his study had taken place in 1880, prior to access to formal education for the masses, a difference would have been identified between the educated middle classes and the limited education of the working classes, within the same economy.

A study by Washington University concluded that a high probability exists that intelligence is more likely to be passed through the female gene than the male gene. The study concluded that a secure emotional bond between mother and child is crucial to growth in some parts of the brain. This is, however, a mix of both nature and nurture; the study goes on to conclude that an emotionally supported child can have a hippocampus up to 10% larger (the part of the brain responsible for memory). I conclude that cultural support post-birth can also influence intelligence.

It is evident that very different Homo sapiens (intelligence wise) co-exist on Earth today, and now could be categorised as differently as Homo sapiens from Neanderthal man, when they co-existed, maybe more so. Not in terms of DNA changes generally, but significant change to the human brain has taken place with apparently little change in DNA. Neanderthals who coexisted with Homo sapiens 200,000 years ago may have had significant DNA differences to Homo sapiens of that time, brain size was similar and intelligence is unlikely to have been significantly different, given that they both competed and survived in similar environments.

A study of genetic structure in human populations by Stanford University highlights the relatively small differences between DNA across different countries and cultures – this signposts to me that little difference exists physically between human DNA in Japan or Chad, for example; average IQ differences are, however, considerable.

Scientists also record that modern humans share considerably more DNA with early Homo sapiens than early Homo sapiens with co-existent Neanderthals. Yet to the observer the behaviours and daily lives of these co-existing species was similar and very displaced from how the vast majority of modern humans live their lives.

Around 70,000 years ago that very close similarity conquering environment changed gradually with the cognitive revolution. Yet although relatively little DNA

CHAPTER 10

differences exist between Homo sapiens of 80,000 to 300,000 years ago as compared to 40,000 years ago, there are major recorded differences in intellect and achievement. The same can be said of Homo sapiens of 40,000 years ago and modern humans. Little DNA changes but substantial changes in application of intellect and achievement.

These differences appear to continue and accelerate in modern humans and their fellow contemporary humans stimulated by the modern phenomenon of humans' ability to change the environment. In just over 200 years the whole dynamic of humanity has changed significantly. If we compare New York, London, Paris and Tokyo, with Bamako, Kinshasa, Bangui, Port Au Prince, Sana'a and N'Djamena – these appear different worlds, yet it isn't that long ago, these cities would have had more comparable priorities for their inhabitants.

The differences: organisation and resources which created wealth, education, structure, better nutrition, healthcare, opportunity and challenge, including mental stimulation. This is driving intellect and ensuring further progress and widening gaps.

DNA with its wide remit may not be measurably changing within and between modern humans. It only has to change a fraction and include the factors which control reasoning within a small part of the frontal lobe which forms part of the Neo-cortex, to enable significant change to become inherited intelligence within the human brain. This, rather than the longer-term effect of physical change through adaption to changing environment by natural selection.

Maybe it's time for science to change how it measures evolutionary progress in humans. The evidence is clear that little change has been measured by defining differences in human DNA, yet substantial differences are obvious in behaviour, intellect and achievement. Measurement will be difficult; however, the work Benyamin has done with the protein FNBP1L is a starting point; it does, however, appear self-evident that significant difference exists, the currency to describe them biologically appears insufficient.

Some Homo sapiens might be evolving into a sub species Homo sapiens (academia) over the long term. Evidence of considerable difference is growing. To be a different species, under scientific definition would need clearly identifiable differences in DNA. I'm not certain that the studies I have seen are absolutely conclusive, but they do indicate a shift influenced by a very small period of time in the context of evolutionary change. Homo sapiens certainly occupy a very broad base when measuring collective intellect needed for economic success.

In summary, the difference in DNA between a modern human with an IQ of 50 compared with one in the 200s will be minor compared with the difference between that of ancient Homo sapiens and Neanderthals. The difference between modern humans is better recorded through actions and behaviours. We can't, however, record

CHAPTER 10

them as different species or sub species because they don't suit scientific definition. For now, we will have to accept the 'broad church' of human intelligence and allow the difference to remain unclassified in evolutionary terms.

I mentioned emotional intelligence earlier – this should not be confused with the emotional brain. The former is far more greatly developed in the modern human brain than other animals. For example, empathy – a human is likely to give a beggar whom it has never met, a sandwich; a chimpanzee would not do the same with food, it is an ingrained survival issue. The emotional brain has, however, existed since early mammals and appears to have developed in a negative way within humans. It can manifest itself as a weakness in human character. The hopefulness of winning the lottery or falling for a scam are embedded, for example.

The UK lottery has around a 14 million to one chance of winning, the top prize around £2-£7million and £2 entry fee. It would need to be entered by an individual for nearly 136,000 years twice a week, a to cover all numerical combinations and still not guarantee a jackpot win. The odds of winning the Euro lottery are many times that figure, yet billions per year is spent by individuals who can't afford to play and maintain a reasonable standard of living at the same time. The emotional brain ruling the practical brain.

When a vulnerable person is telephoned to be told they have won a lottery they don't remember entering and for a release fee of £300 they will receive many tens of thousands of pounds, a surprising amount of people comply. This is a further example of the emotional brain dominating the practical brain. Hopefulness and the wishing of outcomes, such as God, heaven and Santa, for example.

Highly intelligent humans regularly allow dominance of the emotional brain over the practical brain. I conclude this is why many intelligent humans who have merely been told of a human-described God's existence, with no evidence, believe it so unquestioningly. Some 'so called' believers in God have admitted to doing so because there is no risk attached, whereas not believing could be an issue following death. The emotional brain meets the practical brain.

Propaganda preys on the emotional brain and in many cases becomes irreversible. A destroyed German people followed Adolf Hitler to the grave in early May 1945, this because the ideals embedded by propaganda mixed with fear could not be reversed, even within the rubble of a destroyed homeland.

Those humans with low levels of intelligence, driven to a greater extent by survival needs and competitive essentials, may be far less likely to have highly developed emotional intelligence. Emotional intelligence might challenge the impact of the human 'selfish gene' in future humans or a successor species.

The IQ gaps between top scientists, astrophysicists, mathematicians or engineers etc, and humans of

CHAPTER 10

lower-than-average intelligence, are surely as great, if not greater, than that between Homo sapiens and Neanderthal humans when they co-existed on Earth and prior to the cognitive revolution.

As stated earlier, humankind dominates, not because we are stronger, faster or larger, but because we are cleverer. We can now make the same comparisons among certain humans in contemporary society. Levels of education and learning are broadly commensurate with levels of opportunity, wealth and social standing.

In the short term, this current difference could in some cases be explained away by lack of educational opportunity for some social classes, but as I have discussed earlier in this book, we know that constant exposure to changing physical environment changes DNA. Humans are at a very early stage of using DNA differences to identify 'intellect' differences in humans; to my knowledge it can't be done conclusively. DNA is used to identify differences in Genus and Species successfully and accurately. Also, for identifying very slight differences within human DNA in solving crime, inherited medical conditions or predispositions and bloodline lineage, for example.

Lynn B Jorde, Utah University identifies minor genetic differences in modern human DNA but these are in the region of 0.1% or 3 million pairs in 3 billion pairs within Haploid cells. Clearly not enough difference to make distinct physical human species, when we compare what exists between Neanderthals and Homo sapiens. Yet difference in the intellect needed for complex, mechanised economic based skills appears to exist between modern humans, to a significant extent. Lack of opportunity versus continued opportunity generation to generation. Benyamin certainly makes the case for using the FNBP1L protein as a starting point for more investigation in future.

We know some physical differences are confirmed in these slight differences. For example, Nepalese people living at altitude can survive and even flourish at altitude where an average human would perish. DNA over generations has ensured body adaption to an extreme environment. A clear physical adaption to environment.

It would be consistent with evolutionary evidence that constant exposure to complex learning, relatively recently accessed by the human masses, could see abilities inherited through slight, difficult to measure changes. Humans have greater reasoning powers passed down from parents. It is logical and consistent with evolution responsible for levels of reasoning that this could pass slightly modified and improved DNA information from generation to generation. This over time, may become embedded.

Changes necessary in DNA to identify physical differences between species, may be far more obvious and detectable than those needed when compared to significant differences in human intellect. This could explain a noticeable difference in behaviours and performance in such a short time.

CHAPTER 10

Flynn's findings appear to be convincing in persuading us that human intellect is increasing with the benefit of a conducive environment. The average improvement of three IQ points per decade since the 1930s is significant, suggesting a divide between developed and underdeveloped nations is growing. If we take off three points for every decade, we won't be able to go back very far in history. If, for example the average IQ in the 1930s was 90, that would only be 30 decades (300 years) until IQs were theoretically zero.

A preposterous proposal, yet one that clearly indicates a substantial increase in IQ levels since the advent of education for the masses in the late 19th and early 20th centuries in developed countries. Easier access to knowledge and mental stimulation through the internet appears certain to accelerate this and increase this gap in future.

If we could journey back in time to connect with the scholars who wrote the Bible and various religious texts that many people rely on today, we would likely discover that most of them were intellectually lagging, significantly behind our modern standards. Their IQs and emotional intelligence would likely fall significantly below contemporary averages, yet these people of lesser intelligence, with none of the contemporary knowledge of today, continue to exert control over our societies through their outdated work and beliefs.

Not all were unintelligent; highly intelligent people did exist hundreds of years ago, but it is the average IQ that has risen significantly. Galileo, Newton, Aristotle et al are examples of scholars with very high levels of reasoning. Some contradicted religious teachings and many were too afraid to oppose the church, even in a minor way.

Astute religious scholars and others likely existed, understanding that adhering to the church teachings offered safety and improved quality of life, whilst opposing, posed significant risks. Relatively intelligent people would have easily manipulated the masses on behalf of the powerful, as average IQs would likely have been much lower than the averages of developed countries of today.

Alterations in human DNA ultimately represent the impact of parental and ancestral interactions with new environments, adapting to both physical and mental challenges through evolution. Much improved nutrition and opportunities for challenging our brains with complex education are taken up by some but not others, for many varying reasons, many denied them. This factor appears to be accelerating within this apparent and growing, possibly due to evolutionary, as yet unprovable 'brain-reasoning' gap.

Humanity is still at the stage where a child of below average intelligent parents can be nurtured to achieve, if offered the right opportunity. But the vast majority either don't get the opportunity or choose not to take it. Children from countries with low average IQs will likely face increasing challenges in excelling in the future, even if opportunities arise. They will still evolve; however, the challenges and pressures within their environments may continue to diversify from that of developed economies.

As humanity progresses, this dilemma appears to become more prevalent, with developed economies widening the gap between high achievers and those who lack access to opportunities or opt not to seize them over generations. This is likely to lead to significant cultural variations in learning processes.

The proportion of time spent on accumulating resources to survive, personal and family security in third world countries especially, testifies to a different way of life. Wealthier Western democracies offer intricate systems that provide resources, opportunities, education, and pathways to accumulate wealth. These devote more time to leisure pursuits, including studying instead of focusing on more basic survival related matters.

At present third world countries have many individuals with relatively high levels of intelligence, that given opportunity, they too can excel and produce offspring with cultural learning embedded in future generations. However, with less opportunity for mental stimulation, learning and challenge, and developed economies racing ahead, this is likely to maintain and extend any current gaps.

Achievement through funded learning in Westernised high performing economies is already reflected in the top five countries receiving Nobel Laureates: United States, United Kingdom, Germany, France and Sweden. The vast majority of the world's top performing 100 Universities are located in the highly developed Western economies or with a notable amount in the Far Eastern top economies.

This phenomenon is deeply integrated into the educational frameworks, as top schools, teaching hospitals, and educational institutions form commercial and research partnerships. These partnerships attract substantial funding because of their potential for commercial gains and typically occur within the same nations. Most of these countries offer the better jobs, broader internet access, and the freedom to utilise all available learning tools.

This position proves to be self-sustainable and self-reinforcing, as wealth purchases improved education and better education leads to yet greater wealth. In a world where cutting edge technologies are revolutionising human lives so dramatically, this situation will be exacerbated. Capitalism and free markets are relatively new to humankind; it has created great divisions in wealth, those divisions are now likely to be accelerated in other areas of human achievement other than wealth accumulation.

Adding more context to recent evolution strengthens the potential for future human transformation. Up until the late 19th century, about 130 to 160 years ago, which is just a brief moment in evolutionary history, most people in relatively wealthy Western societies struggled for survival.

CHAPTER 10

In London, the capital of the wealthiest nation on Earth, at the time, child mortality rates were high, little or no welfare state existed, work was intermittent, low paid and scarce, and death was a much more regular occurrence within families. Infant mortality and birth rates were both high, a way of life, up to 220 infant deaths per thousand. Child mortality rates for five years-olds and under, as high as 57%.

For 70-100 further years, people struggled with wars and continued lack of opportunity. It is only in recent decades that the social classes have been so large and diverse, with substantial learning opportunity now presented to it, through increased wealth and technology.

After resolving these dilemmas, populations advanced, and Flynn's research shows improvements. When added to contemporary learning methods and enhanced nutrition to promote brain development, this will increase existing gaps. Opportunity exists for some disadvantaged, but it remains to be seen if their personal culture is sufficiently in place to take it within certain sections of society.

Our lineage features numerous splits spanning the last 3.8 billion years. Life evolved from single-celled organisms to multicellular forms, developing into fish and eventually transitioning to land dwellers with lungs. Subsequently, evolution created primates that diversified into apes, eventually resulting in the development of Homo sapiens. It comes as no surprise that we have some initial evidence, albeit small, indicating a further divide, psychologically if not physiologically, following a substantial but uneven advancement in human cognitive abilities, over the last century.

Splits of species are usually gradual and are happening within animals and plants, as a permanent and ongoing feature of life on Earth, albeit slowly. Rapid shifts in environments, like ice ages and thawing periods, have accelerated change among certain species and caused the extinction of others.

DNA is an effective measure in comparing physical differences between species; intelligence differences between the same species is a different challenge. Scientists would not attempt to compare the intelligence differences between two Arctic foxes, the less intelligent would likely have been filtered out over time by natural selection, given that their intelligence would be measured by survival instincts and in life learning. Nature and nurture. Humans with weaknesses as compared with other humans are protected by collective cultures. Natural selection has lost its level of control to human civilisation and collective culture. A good thing, certainly, but it does have the perverse effect of driving up human population and diversity, failing to weed out weakness as it is programmed to do.

Will the rapid advancement in human learning combined with improvements in nutrition and other environmental factors lead to a quicker divergence of the human species

CHAPTER 10

over multiple generations? I believe there is a high chance that it will, and that a split in humankind, if not physically but certainly intellectually, is not a guess, it appears an inevitability. The only remaining question is how fast and how obvious?

To summarise, some humans will outgrow other humans, natural selection will make certain of it. Small indications are already clear in a relatively short time. Humans who endure and challenge their environments will inevitably adapt. Generation after generation studying complex subject matter, for example, and stretching minds, will give future offspring an improved chance of understanding yet higher complexity, guided by their own experiences. The evidence of progression is clear. DNA responsible for reasoning and learning capability and natural selection will have their input. DNA differences may be slight but achievement levels and differences will be clear.

FORCING THE PACE

Human DNA-genes like those of all animals, plants and fungi, are inherited from the generation before, in the case of most animals, from two parents. A mix of genes, we know these have experienced minor change from generation to generation until over thousands of generations significant change and adaption takes place.

Cross breeding of domesticated animals, for example, breeding dogs for different purposes, livestock for increased meat and milk production. In addition, horses for size or speed etc demonstrates how fast some physical changes can be made to adapt animals significantly, to meet new environmental challenges.

In the majority of countries, humans no longer have need to change physically to meet survival needs. Improved and readily available nutrition has helped grow humans in physical size and mental capacity from generation to generation over the last few hundred years. It is already clear that the future evolutionary progress through natural selection and changing of DNA, should be greater in the human mind, rather than in the human body.

Modern athletes and Olympians further the physical achievements of humans but are generally exceptions to the rule, especially in Western societies where record levels of obesity pay testament to lack of need for physical efficiency to survive and thrive. We no longer need to chase down prey when supermarkets deliver high calorie foods to our homes.

Humankind in Africa developed dark skin to shield themselves from the harmful effects of excessive sunlight. As people moved to Europe, their skin colour evolved over time, to be lighter, to ensure sufficient vitamin D production. Those living in colder climates growing more body hair and maintaining that into modern times when, like our appendix, it has become surplus to requirements. An array of clothing has made body hair obsolete.

CHAPTER 10

Natural selection influences this adaptive process across thousands of generations. We are entering an era characterised by information and education, now evolving into the age of human knowledge. Deciphering the human genome, venturing into space, advancing precision weaponry, improving communications, enhancing the performance of medicine, facilitating organ transplantation, advancing engineering technology, and developing Artificial Intelligence. Even teaching dogs to detect life threating diseases and conditions. Cures for many diseases, medical conditions, detection of cancers through simple testing, a greater understanding of the micro-organisms that cause many diseases and how they function. The list continues. These represent a preview sample of what lies ahead.

The 'convenience life' developed for humans by humans, has brought with it a great dichotomy. People can go through life without putting their brains to any major test. Cradle to grave provision through compliance with society's expectation can become a seamless journey throughout infancy, basic education, manual or menial employment, care in old age then death.

Alternatively, people can access a more rigorous and demanding education from a young age. Concentrated and varied learning, specialised schools, cutting edge learning techniques, university and then high-level specialised employment or research disciplines.

These are very contrasting, differently lived lives, with different expectations and outcomes. Some generations of families in Western societies rarely work – a welfare state system to support them has introduced different expectations of life and therefore changed culture. When mechanisation of employment expands this may become more prevalent unless counter measures are introduced.

This separation gap is likely to further increase as wealthy, educated Western societies progress at speed, within societies a capability difference growing as even greater opportunity is afforded to those prepared to grasp it, and can afford it.

Efforts are being made to balance this by ensuring schooling and university opportunities are given to children of lower income, lower-achieving families. This will raise people into the field of greater opportunity, but all too few. It is unlikely to change culture to any significant extent. Remember, evolution is about advantage, not equality.

Aldous Huxley's novel 'Brave New World' presents a scenario where society genetically engineers its citizens to possess a higher level of intelligence. At the level of the potential held by doctors, engineers, university lecturers, and others. This situation poses significant challenges for society because no individuals with lower intelligence levels existed to perform essential tasks like waste management, sewage treatment, or food production and collection. Society requires a diverse array of individuals with varying skills, education levels, and tolerances. Social classes broadly reflect these differences and become a blueprint for expectation.

CHAPTER 10

When historians look back at our age, they may conclude that we are in an age of coexistence and already, split into more than one human type. This statement may appear to echo far-right political views; however, it does not show bias based on nationality, race, creed, or politics. It simply encapsulates the impact of education and learning opportunities, as these influence culture across generations. For example, children of doctors being more likely to become doctors than those of humans following other occupations.

Unaddressed, with lack of learning opportunities, this gap will keep widening, making it more likely that learning complex topics becomes ingrained in our culture and even more embedded in our genetic makeup. Although no historical studies fully support this claim, humans are in unchartered territory when it comes to future evolution as influenced by contemporary evolutionary pressures for change. Life on Earth has never experienced anything as rapid as the rise of modern humans and the change effected over a very short time. It has been proven that influences such as drug dependency can be introduced genetically, from father to child – could it then be possible to pass increased intelligence and a propensity for problem solving over a few generations?

Humans have been increasing in intellect over tens of thousands of years. The evidence is that modern humans are accelerating this, each generation achieving more than the next.

We know DNA changes over time, significant physiological change can take thousands of generations, the evidence is inconclusive regarding psychological change. The drugs dependency example, used earlier, demonstrates that 'in brain need' for chemical stimulus can pass quickly, so why not propensity to modern learning at a slower yet relatively rapid rate when compared to historical change? Studies on DNA demonstrates slow change and are not reflective of huge intelligence and achievement differences across the Homo sapiens categorised species, over time.

In our modern information and learning age, this dynamic will accelerate change. A wealth of knowledge exists, generated by educational institutions like universities and international organisations, as well as governments and agencies such as NASA through extensive research.

Those who take advantage of these opportunities will not only benefit themselves but also may provide their future offspring with a substantial competitive edge. It isn't necessarily information itself, but the stretching and testing of the human brain which will increase capacity and flexibility of thought. In time, this is likely to influence the chemical structure, makeup and ability to influence from down generations, within human reasoning DNA.

Passed from generation to generation, very minor DNA modifications could be stimulated to change and influence future generations. Individuals with a higher capacity to learn will enjoy an even greater advantage. Those higher achievers will then go on to further advance research, learning and populating information streams which should 'hook' other developing human brains.

When I was younger, knowledge was gained at school, reading or laboriously searching in a library, the occasional TV documentary; now it is at the touch of a few keys on a keyboard or phone in a search engine, which opens 'worlds' of knowledge in seconds, in addition to a plentiful supply of made up nonsense.

RELIGION AND EARLY USE OF HUMAN ADVANTAGE OVER HUMANS

Numerous examples illustrate how the evolved human brain has developed advantages via enhanced reasoning. Humans have achieved remarkable progress in many fields, but today, advancements in weapons and the lingering desire to employ them continue to adversely impact humanity. People have employed their reasoning skills, to achieve advantages in warfare since the very first human conflicts.

Human physical natural selection has been evolving and continuing throughout our existence on Earth. As hunter-gatherers, Homo sapiens relied heavily on a balance between mental acuity and physical strength for survival and to gain an edge over their environment. Gaining more food by designing better weapons and the ability to defend against aggression. A 60kg human with a spear likely to be a match for an unarmed 80kg human. Utilising intelligence instead of physical strength.

Human intellect and its evolution in leveraging weaponry and strategy originated many years ago. This could partially shed light on why bulky Neanderthals failed to compete with Homo sapiens during the struggle for survival. In times of famine or extreme cold when resources were scarce, Homo sapiens likely outperformed Neanderthals due to their superior reasoning and communication abilities.

Neanderthals possessed a brain size comparable to that of early Homo sapiens; however, differences likely stemmed from the size of the frontal lobe of the neocortex, which plays a critical role in reasoning. Since the cognitive revolution began around 70,000 years ago, this difference likely provided Homo sapiens with notable advantages in hunting and combat scenarios, leading to improved chances of survival.

The cognitive revolution and the development of civilisation transformed the dynamics of evolution and natural selection for humankind. Previously, factors such as strength, speed, agility, and hunting skills enhanced the chances of survival. Over time, these physical

CHAPTER 10

attributes gave way to the importance of reasoning capabilities.

Reasoning can, for instance, recruit armies by promoting the advantages of victory over the risks that must be confronted. It might also devise effective battle strategies, also harness intelligence to build strength and secure a competitive edge.

Early recorded battles between humans showcased how superior strategy that stems from effective reasoning can enhance survival techniques. As mentioned earlier, Alexander the Great fought in the Battle of Issus in 333 BC. Macedonian forces were outnumbered by the Persians by 1.8 to one according to modern sources but by as many as eight to one according to ancient texts.

Alexander compensated for his numerical disadvantage with his strategic skills. His use of phalanx infantry as a bulwark against Persian foot soldiers, and his inventive use of cavalry as a hammer blow to the Persian flanks, this overwhelmed the enemy for a decisive victory.

Warfare has experienced transitions for centuries, particularly accelerating in the last 80 years with nuclear weapons dramatically altering the conflict landscape, followed by conventional combat from distance. No longer are wars fought on strength of numbers alone, although that remains a factor, but is of less, and diminishing importance. Modern weaponry takes brain not brawn to develop and perfect. Millions can be killed in a single action regardless of how fast, strong or skilled individuals may be.

A handful of scientists defeated the Japanese nation in 1945, with a few years of development work and less than two weeks after delivery of that completed work. Four years of combat in the air, sea and hand to hand attritional warfare had given the allies an advantage, but possibly a further 18 months plus hundreds of thousands of lives would have been lost, in concluding the Far-East war.

Japan would have inevitably lost but at huge and senseless cost in terms of civilians, service personnel and essential infrastructure. A country so large would have taken decades to fully recover after their young men senselessly sacrificed their lives, delaying the inevitable. By mid-1945, 95% of Japanese soldiers preferred death over surrender, and this extreme belief showed no signs of abating.

Leaving aside nuclear weapons, handheld weapons fired out of range of tank-firing capability and range can now destroy a tank battalion in under an hour. An individual controlling a drone from halfway around the world can cause thousands of deaths and still pick up their children from school, within the same day.

At some point in our history, advanced reasoning took precedence over physical strength or fitness for survival. We now live in a reality filled with nuclear weapons,

CHAPTER 10

human-induced climate change, and lab-engineered organisms that pose a risk to humanity's survival. A small group of individuals could end the lives of millions at a mere push of a button. We could be about to snatch defeat from evolution's 'jaws of victory'.

DOMINANT YET TEMPORARY

To understand where humans are heading, we need the past to inform us. The only information we have to analyse is in the past. Humankind continues to evolve and will ultimately diverge on the evolutionary tree, similar to how hominins diverged from chimpanzees, seven million years ago and up to Neanderthals around 430,000 to 700,000 years ago. Homo sapiens of 200,000 to 300,000 years ago seem very different from modern day humans, yet science continues to categorise them as the same species. An Astrophysicist engaging in dialogue with Homo sapiens of 300,000 years ago would not appear to be the same species. Some scientists estimate Neanderthals arriving as long ago as 750,000 years and Homo sapiens as late as 200,000 years ago.

Humans possess embedded DNA 'instinctive knowledge' along with experiential learning, which resulted in enhanced problem-solving abilities. This advancement ultimately facilitated the rise of farming, community organisation, and the establishment of strength and security through social connection.

Humans have established themselves as the dominant species on Earth, primarily due to their advanced reasoning skills and the physical advantage of having two opposable thumbs, which enable intricate and skilled manipulations. This domination is likely to continue for the foreseeable future.

But this situation was not the case 100,000 years ago and probably may not be the situation in another 100,000 years. Homo sapiens are likely to dominate Earth for a fraction of the nearly two billion years that unicellular creatures dominated it. Humans will not 'out survive' unicellular organisms who will be the true 'winner' of evolution and survival when all life on Earth finally becomes extinct. They will have survived considerably more than half the life of our solar system.

The rise of agriculture eliminated the necessity of being a skilled hunter for survival. A further example of how the development of the human brain gained advantage over speed, fitness and strength in the game of survival. Planning and delivering food supply for the future rather than reliance on a successful hunt or forage every few days.

However, as discussed in other sections of this book, transitioning was challenging for humans because their dependency shifted from hunting for food to cultivating it and raising livestock, while also facing the threats of famine and crop failure. Farming enabled many more humans to live, but sometimes in relative misery, and

CHAPTER 10

many more died, the overall result being a population growth, with a heavy price.

Civilisation and the human groups that created it led to a hierarchy of survival. In relatively modern Homo Sapiens, this hierarchy of survival manifested itself in 'leaders', kings, emperors, sultans, holy men, self-declared through wars or gods. Power grew with numbers raising chances of survival significantly and protecting lifestyle. This was reinforced by fear of the unknown, by use of religious context.

When leaders, kings, emperors, and holy figures introduced God as a factor, many claimed divine anointment and approval. Christians will be familiar with the 'hell and damnation' threat for non-compliance. All for control of human over human, some with good intention, some not. A survival and thriving strategy, for those who finely honed this state of control.

Farming fostered cooperation and the development of communities, which eventually gave rise to hierarchies and the early foundations of politics and religion. Religion stands out as one of the most intriguing creations arising from the human neo-cortex's capacity to reason and make sense of the unseen or unprovable. It also satisfies a fundamental need, stemming from the innate curiosity of a questioning mind.

The questioning human brain had its answer, the reason for the unexplainable was God and 'his' will.

Today, and in future, that questioning brain has become a challenging brain, one which seeks evidence, considers emerging information and alternatives, beholds weakness, naivety and lack of knowledge in ancient texts.

Religion has influenced humanity and civilisation more profoundly than any other factor created by the human mind. Yet only for about 6000 years that we know of, in its modern interpretation, or 2% of Homo Sapiens' time on Earth. Organised religion with places of worship is a relatively recent addition. The inability to challenge an invisible omnipotent God being a master stroke from those in power, in their efforts to retain and extend power. The human brain proves to be highly susceptible to manipulation, as we have observed.

Greek, Roman, Persian and other gods from fairly recent human history died in human brains, when the last of the humans who believed in them died. The type of Gods the human brain creates seem to have depended on the purpose. Power and awareness seem to be very strong themes in modern religions, an 'all seeing all powerful, invisible God' appears to be popular within the omnipotent, monotheistic God religions.

Polytheistic religions, such as those of the Greeks, Romans, Norse, and even modern Hinduism, often focused on specialist gods who addressed survival and cultural needs, including aspects like war, the sea, spirituality, love, the seasons, harvests, animals, wealth, and health.

CHAPTER 10

Humans in different times and environments rationalised the embryo of all these religions, then passed it to the next human generation and their receptive, pliable, needy brain, and so it went on. Additions made along the way by scholars and religious writers, no doubt dismissing the concepts and ideas that didn't suit the religious decision makers of the time.

Given there have been so many different human-described gods, the balance of probability is that none of these exist. They only exist whilst the human brains that conceptualise them exist. As we have discussed, all believers in Roman, Greek, Norse gods et al are dead, in the minds of worshippers, whom are dead, therefore their gods are dead.

The definition of a current human-described god appears to be one that is worshipped and followed by many. Venus, Thor, Neptune and Odin no longer qualify, but long ago they were as all-consuming to humans as the gods of today, maybe more so.

I wonder how Albert Einstein would have been received by those ancient holy men who wrote religious texts, the Bible, or other religious texts? Explaining how gravity bends spacetime and expressing the relationship between mass and gravity. Or a modern-day scientist explaining that beyond the stars is not heaven but hundreds of billions of galaxies billions of trillions of stars. The violence and chaos in the Universe resemble a religious description of hell, and is unlike perceived heaven.

Einstein and his fellow scientists would have very likely been sentenced to death if Galileo's experience is an indicator, having stated that the Earth orbits the Sun and is not the centre of creation. Further, that the Earth is a tiny piece of rock in a sea of almost infinite stars and planets – calling it insignificant in terms of the size of our Universe would be overstating the Earth's significance. Religious statements regarding the significance of the Earth being proven to be poor guesses. The Earth is significant to humans and life on it, not to any other part of our Universe.

The history of religion offers insights into the future of faith. We can understand how a religion can die through wars and the slaying or enslavement of its believers, but how will modern, individual religions die? The answer from history is either from war but more likely from tolerance of other religions.

Later tolerance and then acceptance of Christianity, led to Roman Gods and religions dying through that tolerance of Christianity. The truth about the Romans is that they allowed their conquered peoples to believe in their own gods, providing they paid homage and respected Roman Gods. Christians would only accept their own god and generally refused to comply.

The Romans viewed Christians as atheists because of their refusal to worship idols. This separation and stubborn resistance led to incompetent Roman emperors such as Nero, singling out and persecuting Christians as easy targets, on which to blame the failure of political

CHAPTER 10

decisions, in a not dissimilar way to the Nazis' blaming of Jews for economic failures of the 1920s and 1930s.

Judaism influenced Christianity regarding the concept of a single god. Christianity was not turning away from God(s), for another, in the way that Romans swapped idols for a single Christian God. It was an adoption of an existing god, injecting different beliefs and rituals to worship 'him'. The Old Testament was written about a Jewish God as Christianity did not exist, for the period it covered. The Old Testament is derived from the 24 sacred texts of the Jewish Tanakh.

Christians have claimed God in this period of pre-Christianity, as a Christian God; also Abraham and Noah are Jews, their asserted deeds claimed by the Christian Bible. All prophets served a single God, as does the Prophet Muhammad, as viewed by Islam. Christianity appears to be a sect of Judaism, and according to the Jewish Encyclopaedia, Kaufmann Kohler, Christianity considered itself so, for a substantial period after the death of Christ.

The act of Baptism and conversion to Christianity evolved from the Jewish act of Mikvah, called the Tevilah, immersion in a natural water source, as an act of purification, this had been a Hebrew ceremony before the birth of Jesus Christ or John-the-Baptist.

Many similarities existed between the two faiths, yet the human brains, being an easily influenced lump of flesh, found division and hatred in difference, influenced by superstition and myth. Persecuting the Jews for being complicit in killing a Jew. How many Jewish people have Christians murdered? Hatred through difference has been a theme of humanity for since before it split from primates, its brain has become more easily influenced, religion has exacerbated this, the evidence is compelling.

Apes and other animals kill each other for food, territory, breeding rights, survival and other procreation reasons, all linked to continued existence of them and their DNA, instincts deep within their DNA. Humans kill each other for differences of opinion, on behalf of a phenomenon, on balance of probability, unlikely to exist.

For 280 years after the death of Jesus Christ, the Christian religion was a minority religion within the Roman Empire. The Roman Empire had been split into differently ruled parts. Constantine, the leader of the Eastern Empire, defeated Maxentius to unify various regions of the Empire.

The two armies met, historians believe, at Milvian Bridge. Constantine with an army of between 20,000 to 40,000, defeated a 25,000 to 100,000 strong army of Maxentius, depending on whose account of the battle is used. Constantine believed he experienced divine intervention after overcoming the overwhelming odds. Constantine had converted to Christianity and attributed his victory to the Christian God.

CHAPTER 10

Some historians argue that during Constantine's reign, people began to view Jesus Christ as a deity and the Son of God. Before this era, many regarded him as a prophet or messenger of God, similar to the recognition accorded to the prophet Muhammad, within Islam.

Promoting a deceased hero seems to offer significant benefits. I visited Cuba in 2004 and quickly learned that Che Guevara was a God-like hero, keenly promoted as such by the communist Castro government. Had Guevara been alive, authorities would likely have stopped him from attaining such a status, as it could have threatened Castro's own position. Similarly, had Jesus Christ had been living in the time of Constantine, he could have been viewed as a threat to the status of the emperor.

If a modern-day Muslim claimed to be a new Prophet, or a Christian asserted they are the son of God, they would likely face disdain, possibly viewed as mentally unstable. If populations took them seriously, they would likely experience some form of oppression from others. Muslims believe Muhammad is the last prophet from God and I assume any person claiming to usurp that position would not be taken seriously, indeed likely to be treated with anything from disdain to being perceived as committing heresy.

Numerous individuals have claimed to be the new 'latest son of God' or a prophet of God, but as education has improved, people's gullibility has considerably decreased, none have been taken seriously by more than a handful of people.

Following the Battle of Milvian Bridge 312 AD and then securing the empire in 324 AD, it appears that Constantine wanted to make the strong statement and pay homage to his new Christian God. There was obviously huge advantage to be gained in this account of Christ's new status, when trying to convert followers of other religions to Christianity. Baptism being the threshold for change.

Many battles had been won against overwhelming odds. Mars, the Roman God of War, was given credit for the battle of Watling Street, which had occurred 250 years earlier, when approximately 10-11,000 Romans defeated between 100,000 and 230,000 Britons, depending on different accounts of the battle. Alexander the Great's defeat of the mighty Persian army at the battle of Issus in 333 BC also alleged testimony to Ares the Greek God of War. History, of course, written by the victors.

A significant shift in religion, when it occurs, often occurs after a lost war or another major event. The Milvian Bridge event launched Christianity into a mainstream religion embraced by many, eventually making it the most widespread and popular religious faith on Earth.

Under the Romans between the first and the fourth centuries Christianity had grown, albeit slowly, both inside and outside the Roman Empire from the time of Jesus Christ, Constantine giving it a major boost with his conversion during the early fourth century.

CHAPTER 10

By the conclusion of the first century, historians estimate that Christianity had fewer than 10,000 followers. By the year 300 AD, the number of Christians had grown to approximately six million. By 350 AD, this total surged to an estimated 33 million, which constituted more than half of the Roman Empire's overall population, of around 60 million.

The catalysts being the support of the most powerful man in the world and battles won in 312 AD and 324 AD. The truth appears to be that battles win gods, rather than the reverse. In a mere 33 years, Constantine had increased the following of the Christian religion by 550 per cent and added in excess of 26 million followers, assisted by considerable violence and human deaths.

Many more Christians being directly attributable to Constantine in a 27-year-reign, than Jesus Christ in 33 years of life; however, all Christians are, of course, either directly or indirectly attributable to Christ. But 280 years following his death, just six million followers were known to exist, a minority religion in the Roman Empire.

Then 33 million Christian followers by 350 AD ensured a solid 'Beach Head' with which to spread the word of Christianity, eventually worldwide. Christianity has increased more than 70 times worldwide in the intervening 1700 years.

More than half of the Roman populace and the victors identified as Christians, so the narrative would have shifted considerably if only 10% had been Christian prior to the Milvian Bridge and faced a loss; persecution would likely have followed, due to Maxentius's inevitable desire for retaliation.

Under those conditions, Christianity would be unlikely to have grown into a major global religion. The final battle in any religious conflict carries significant weight, since it often determines the victors of the war and who possesses the true God. The wider introduction of a any specific god and religion could cost many lives and great human suffering. The introduction of Christianity to South and Central America by the Spanish bears lasting testimony.

After Constantine's triumphs in 312 and 324 AD, Christian education and places of worship flourished in the Christian Roman empire. Over the centuries, Christian religion became less tolerant of non-believers and alternative views. This guaranteed the enduring presence and supremacy of Christianity in its practised locations.

Christianity has exhibited semi-tolerance at best, or more usually intolerance towards other religions for more than 1500 years while flourishing in Europe. Over the past 150 years, particularly in the last 30 years, the Christian religion has exhibited greater tolerance while Western society has embraced multiculturalism.

I hold no opinion on whether the decline of one religion in favour of another is beneficial or detrimental; I simply observe and wish I had been born in a future era

CHAPTER 10

devoid of religion, ruled by logic, facts, intelligence and evidence, not superstition.

The step change in the Christian religion has been a decrease in those practising its religion. The timing of this decline aligns with the information age of the internet, which promotes the exchange of ideas and discussions in free and accessible societies.

The human brain has now met science and discovery, and in many cases these discoveries are not consistent with religious teachings. Religious believers often attempt to force factual scientific discoveries to align with their sacred texts and teachings. The internet has been commonly available since the early 1990s, information and knowledge available has increased, therefore tapered during that time. Behold the change in people's views.

In contrast to earlier approaches, this strategy shifted from denying the validity of discoveries by figures like Copernicus, Galileo, Newton, and Darwin. A clear sign that religion has had to take science and discovery seriously, or look backward and outdated. Using threats of excommunication or the killing of innocent people can no longer be considered an acceptable strategy in modern, democratic society.

We know the Earth is more than 6600 years old, we know it has a history of over 4.6 billion years, we know humankind is next to insignificant compared to the expanse of the Universe, the time it has existed, and the time it is likely to exist. Consequently, we understand that the authors of religious texts throughout the centuries were speculating, as they lacked the scientific information available today.

It is with unbelievable arrogance and ignorance that humans still believe the Universe or even the Earth was created with humans as 'the chosen ones'. Considering the evidence available, it amounts to complete nonsense.

'Creationism, of course, is nothing more than thinly disguised religion masquerading as science, in order to attempt an end run around the US Constitution's First Amendment prohibition on the Government establishment of religion'. Richard Dawkins

Humanity should realise and accept its far more minor and latter part, in the story of the long story of Earth. 'Johnny Come Lately' now destroying it and therefore we are having a great impact on it in the present. The Earth will continue to exist long after humanity has become extinct, and it has demonstrated remarkable recovery abilities over time. The five extinctions in the last 500 million years paying testimony to this.

Earth will remain for a significantly longer duration than the gods conceived in human minds, yet it cannot exceed the lifespan of electromagnetism, gravity, quasars, or black holes – the actual and identifiable powers of the Universe. These phenomena have existed for billions of years, yet Homo sapiens and their gods are highly

CHAPTER 10

unlikely to last a million years, which is merely a blink of an eye in the timeline of the Universe's past and future.

Those inaccuracies, of the ancient religious text writers were understandable, given the times when they were written; what seems incredible is that this misinformation persists today. A few centuries ago, access to religious texts often connected strongly with social status; if you could enhance or invent a narrative that glorified God, it significantly elevated your standing with both ecclesiastical and political authorities. Christianity faces significant challenges, and I understand the reasons behind this.

Christianity, ironically, faces greater threats as it adopts a more tolerant attitude toward other religions. In many countries around the world, the religious and political leadership have little tolerance for the growth of religious difference and cultural integration, preferring to protect their own beliefs from outside influences – many Muslim countries, for example.

Modern liberalism and inclusion throughout Europe, has, and is, embedding tolerance of other religions into Christian society. The Christian church would not have chosen this situation, clearly indicating that the influence of liberal politics has surpassed that of traditional religious will, in Westernised societies.

Many individuals in Western countries can now deviate from religious teachings and attendance at places of worship due to increased freedom of choice, a feat that would have been difficult or impossible in the past. In many Christian countries, individuals faced ostracism for not engaging in religious practices such as attending worship services or studying holy texts. Many people in religious autocratic countries experience repercussions for not participating in group worship or adhering to religious doctrines.

In countries that impose penalties for not attending worship, such as ostracism, the attendance percentages remain noticeably higher. The conclusion is that a significant level of religious following exists because of limited or no access or exposure to information offering an alternative view, and/ or the fear of consequences of expressing an alternative view.

Liberalism enabled me to write this book freely, without fear of oppression. I would have been executed as a heretic only a few hundred years ago, and currently, in less liberal countries. In a Western world that embraces freedom of information and tolerates various religions as well as non-religious perspectives, Christianity must come to terms with the accompanying decline. Other religions enjoy protection in their countries of origin, showing intolerance, and they are likely to expand and thrive, from that base, within accessible freedoms abroad.

Religion tends to be less prevalent per thousand people in many countries with higher educational standards among the general population, or where education does

CHAPTER 10

not mandate religious study. In nations where religion is presented as an indisputable truth and where individuals must either believe and follow it or face repercussions, people tend to hold strong beliefs in religion, and they are more likely to follow its teachings.

How long will this situation continue? The hold religion has on humanity is loosening, especially in Westernised culture. Many countries no longer create laws based on religion; instead, they base their laws on moral principles. Although some of these principles originate from religious texts, lawmakers examine their compatibility with freedom of speech and human rights, moving beyond religious doctrines.

Whilst access to the human mind is controlled, then religion will continue to flourish, freedom of access to information, especially science and uncensored learning will lessen the grip of religion on human minds, more so when more people are given access. Humans are becoming more intelligent, religion is a convincing story; however, it is far less convincing for those with knowledge of alternative views.

Church attendance has declined, and the increase in non-religious funerals clearly indicates a weakening hold of traditional beliefs in liberal countries that promote freedom of learning and expression. Faith and spirituality have longstanding significance for the human brain and often connect to a concept of God. For thousands of years, before people defined gods and religions, faith and spirituality undoubtedly resided within the human mind.

Ancient cave paintings of animals, some dating back 45,000 years in Indonesia and featuring similar age examples in various countries worldwide, demonstrate this form of non-religious spirituality. Animals being held in high esteem, possibly awe, by ancient humans for power, strength, provision of food and providing evidence of what humans saw as important enough to visualise and record, in a similar way to the symbols later created in early religions.

When the first upright human looked into the night sky, he or she must have been filled with wonderment and experienced the brain stimulus as modern humans that we refer to as faith or spirituality. I have faith and spirituality, but I know they are only cerebral stimuli caused by changing chemical levels in my brain, but they give me wonderment, maybe the same effect that drugs have on an addict.

As the human brain develops further and new discoveries are made, it seems certain to reject unknowns and superstition for the knowns of science and discovery. Throughout history, humans have replaced one set of deities with another, as seen when Roman, Greek, and Norse gods transitioned to Christian figures in just a few generations, similar to transformations recorded on other continents. In future, it is likely that gods will be replaced by forces that imitate the impact that human-conceived gods currently make.

CHAPTER 10

Gravity, spacetime, electromagnetism, strong nuclear power and black holes, all of which have played a key role in the Universe and therefore the story of life from the very beginning. They have been with our Universe for most of its 13.8 billion years; gods as defined by modern humans, are likely have existed for considerably less than 10,000 years.

Humans, and what comes after, are unlikely to worship gods, worship of the non-existent will be for the superstitious, but humans will understand the essential role the great powers play in creating and driving everything.

Gods and worship will be known as a period of humanity with a start and an end, similar to the way we recall the Egyptian, Greek and Roman empires. They rose, they fell over hundreds of years; modern human religion is likely to have risen and fallen over 8,000 to 12,000 years, maybe less.

Children are likely to discuss the great powers of our Universe in awe in school classes, in the distant future, in the same way they discuss religion, Santa Claus and pop stars today. Their IQs, emotional intelligence and capacity to learn, way beyond our current imagination.

Humans will be very temporary within the life of the Universe and of that on Earth – 300,000 years of Homo sapiens is a comparative blink of the eye.

Human intelligence levels will increase substantially to enable children to understand complexity, that many adults fail to understand at present. Eight-year-old children being able to pass a physics degree in 12 months. They will later accelerate human achievement yet more. It isn't far-fetched, but a matter of time. The difference in the level of maths and science my children were expected to learn throughout their schooling demonstrated how much more challenging schooling was in the 1990s and 2000s, when compared with the 1960s and 1970s. Further supporting Flynn's findings on rising IQ levels.

A rapid rate of growth in complexity and challenge in just three decades. My children were studying the same level of maths at ages eight that I reached by age ten. Children are likely to achieve levels of education earlier and earlier as standards improve; these will foster ongoing advancements in human discovery and creativity. It will be embedded in a small part of our DNA, possibly with huge effect.

We will learn significantly more about the Universe. The laws of physics explaining most but not all, then a bit more and a bit more until humanity understands far more about how the Universe works. Let's set aside the idea of a divine intelligence and examine conclusions that have more scientific basis.

The cracking of the human genome code was a huge challenge which humanity has met; discovering how the Universe works is getting closer, and to have concluded that it was developed through a seven-day event, including a day off, would have been a travesty to learning and an insult to the known gods.

CHAPTER 10

The church and politicians of 400-500 years ago would have prevented the study of the Universe, or at least outlawed its conclusions as they did with Galileo. Countries and their religions still exist around the world that would prevent progress in this area.

I am fortunate to live in a society that witnesses and accepts the wonders of our Universe and creation, as explained through science. In my view, religion has proven retrograde to scientific discovery, and it should carry a degree of shame.

Texts written by individuals with little information between 500 and 3500 years ago still dominate some countries' laws and behaviours in the 21st Century. Where faith, morality and behaviours are its aim, it is surely a good thing, but these elements need separated and disseminated along with a believable format of introduction to human learning. Where information has been disseminated for gain and human domination over humans, this will be lessened as the human brain develops and is no longer manipulated by religion.

Faith and spirituality will continue to exist when current religions and gods meet the same fate as Roman Gods; my hope is that the forces of the Universe, the known Gods, are given the credit they deserve.

Developing my views over time has left me without an all-powerful human-described God, and many will find that sad. To me, it is not an issue, as I have long since lost trust in those I once trusted, though my affection remains, they meant well. I have huge compensation, in the wonders of the Universe and the story of evolution of life on Earth.

Appreciating the awesome power of objects billions of light years away and billions of years ago, that were essential for every form of life on our planet, and where it may exist in our Universe. The creation as explained by science is, I believe, evidence based – we may not have all the evidence, but the story is at least a convincing one to my questioning mind.

The evidence may not be what I would have wanted as a six-year-old – a cosy heaven and eternal life may have been preferable – but senior to that in my contemporary mind is the search for the truth. Religious commentary appears light years from that truth.

Many mysteries removed by science, replaced by more mysteries which challenge my mind. If enough time remains with sufficient development in human brains, or what succeeds them, more questions about our journey and that of the Universe will be answered.

I am as sorry that I can't be with those of you yet to live, to share them, as I'm sure Einstein would be, about missing out on the discoveries of the Hubble and James Webb telescopes. I have lived in a great age of discovery, and I wouldn't change it, but that doesn't stop me being very envious of those of you yet to make life's journey.

CHAPTER 10

You are in oblivion now and will swop that oblivion for life, consciousness and realisation, sometime after I greet oblivion like a long-lost friend. I will return to my natural state.

We have the answers that legends like Einstein gave us, but also have knowledge that they never had. We should feel so privileged that we live in a generation that can look across time and space 'having stood on the shoulders of Giants'.

*

References, sources and recommended reading or research, Chapter 10

The Impact of Social Classes on Education Opportunities in Different Countries – Marteusz Brodowicz

School Leavers, England's history of School Leaving Age – Leavershoodiescompany.co.uk

If you live in these countries you're breathing the most polluted air in Europe – Gabriela Galvin, Euro News, Jan 2025

Triune Brain – Paul MacLean, Science Direct

Genetic Structure of Human Populations – Noah A Rosenberg et al, Stanford University

Genetic Variation and Human Evolution – Lynn B Jorde Ph.D, Department of Human Genetics, University of Utah, School of Medicine

Has Human Evolution Stopped – Alan R Templeton, National Library of Medicine

Is there still evolution in the human population? – Super Nature Link

The Human Brain in the Modern World – Charlotte Kume-Holland

CHAPTER 10

The 10,000 years explosion, how civilization accelerated human evolution – Gregory Cochran and Henry Harpenden

Are We Stalled Part Way Through A Major Evolutionary Transition from Individual To Group? – Stephen C Stearns, Department of Ecology and Evolutionary Biology, Yale University

Thank your parents if you're smart: Up to 40% of a child's intelligence is inherited, researches claim – Dr Beben Benyamin, University of Queensland

Children inherit their intelligence form their mother not their father – Charlotte England, reporter, Independent Newspaper, Science section

Has humanity reached 'peak intelligence' – David Robson, BBC, Deep Civilisation, Brain

Cracking the Human Genome – Kevin Davies, John Hopkins University Press

Ancient Gods and Goddesses from cultures around the world – Matthew Jones, History Cooperative

The Scandinavian System (or why atheism is a belief system) – Dr Andy Bannister, Third Space

Christianity in relation to Judaism – Kaufmann Kohler, Jewish Encyclopedia

Assessing intellectual abilities of asylum seekers – Simon Whittaker, National Library of Medicine, National Center for Biotechnology Information

Education standards in historical perspective – Richard Aldrich, Learforlife.medium

Education: historical statistic – Paul Bolton, House of Commons Library, standard note SN/SG/4252

How important is education for economic growth? The relationship between education and economic growth – Anna Sudderth, xqsuperschool.org

Humans continue to evolve with the emergence of new genes – Dan Gray, Medical News Today

Sophie Scholl and the White Rose – The National World War II Museum, New Orleans (February 2020)

ACKNOWLEDGEMENTS, SPECIAL MENTIONS AND INFORMATION SOURCES

- *Joseph LeDoux*, Neuroscientist and Author
- *David Eagleman*, Neuroscientist and Author
- *David Goleman*, Psychologist and Author
- *Yuval Noah Harari*, Historian and Author
- *James Flynn*, Philosopher
- *Frank Drake*, Astronomer

My Inspirations

- *My family and friends*
- *Albert Einstein*, Theoretical Physicist
- *Galileo Galilei*, Astronomer
- *Edwin Hubble*, Astronomer
- *Isaac Newton*, Scientist and Master of the Royal Mint
- *Johannes Kepler*, German Astronomer, Mathematician, Astrologer and Natural Philosopher
- *Charles Darwin*, Naturalist and Geologist
- *Winston Churchill*, Prime Minister of Great Britain 1941-1945 & 1951 to 1955

- *Dr Martin Luther King junior,* American Civil Rights Activist
- *Josef Gabčik and Jan Kubiš,* assassins of Reinhard Heydrich, Acting Protector of Bohemia and Moravia, representative of Nazi Germany
- *Sophie Scholl (9th May 1921 to 22nd February 1943),* White Rose anti-Nazi resistance, beheaded for distributing anti-Nazi literature at the age of just 22.
- *'Richard Dawkins,* evolutionary biologist, zoologist, science communicator and author
- *Michelle Thaller,* Astronomer
- *Hakeem Oluseyi,* Astrophysicist
- *Phil Plait,* Astronomer
- *Paul M Sutter,* Astrophysicist
- *Chiara Mingarelli,* Astrophysicist
- *Joseph LeDoux,* Neuroscientist and Author

How the Universe Works Additional Contributions from:

Alex Filippenko	Astrophysicist
Hakeem Oluseyi	Astrophysicist
Grant Tremblay	Astrophysicist
Micelle Thaller	Astronomer
Amber Straughn	Astrophysicist
Phil Plait	Astronomer
Paul M Sutter	Astrophysicist
Katie Bouman	Data Scientist
Shep Doeleman	Astronomer
Chiara Mingarelli	Astrophysicist

ACKNOWLEDGEMENTS

Katherine Freese	Theoretical Physicist
Sean Carroll	Theoretical Physicist
Moogega Stricker	Planetary Protection Engineer
Alessandra Pacini	Physicist
Sarah Stewart	Planetary Scientist
Kevin Walsh	Astronomer
Andrew Pontzen	Cosmologist
Jessica Esquivel	Particle Physicist
Philip Hopkins	Theoretical Astrophysicist
Dan Durda	Planetary Scientist
Jani Radebaugh	Planetary Scientist
Kevin Walsh	Astronomer
James Bullock	Astrophysicist
Konstantin Batygin	Planetary Astrophysicist
Lewis Dartnell	Astrobiologist
Dan Durda	Planetary Scientist
Jani Radebaugh	Planetary Scientist
Prof Michlo Kaku	Theoretical Physicist
Prof Carlos Frenk	Cosmologist
Prof David Spergel	Theoretical Physicist